INNATE IMMUNITY TO PULMONARY INFECTION

The Novartis Foundation is an international scientific and educational charity (UK Registered Charity No. 313574). Known until September 1997 as the Ciba Foundation, it was established in 1947 by the CIBA company of Basle, which merged with Sandoz in 1996, to form Novartis. The Foundation operates independently in London under English trust law. It was formally opened on 22 June 1949.

The Foundation promotes the study and general knowledge of science and in particular encourages international co-operation in scientific research. To this end, it organizes internationally acclaimed meetings (typically eight symposia and allied open meetings and 15–20 discussion meetings each year) and publishes eight books per year featuring the presented papers and discussions from the symposia. Although primarily an operational rather than a grant-making foundation, it awards bursaries to young scientists to attend the symposia and afterwards work with one of the other participants.

The Foundation's headquarters at 41 Portland Place, London W1B 1BN, provide library facilities, open to graduates in science and allied disciplines. Media relations are fostered by regular press conferences and by articles prepared by the Foundation's Science Writer in Residence. The Foundation offers accommodation and meeting facilities to visiting scientists and their societies.

Information on all Foundation activities can be found at http://www.novartisfound.org.uk

Novartis Foundation Symposium 279

INNATE IMMUNITY TO PULMONARY INFECTION

John Wiley & Sons, Ltd

Published in 2006 by John Wiley & Sons Ltd,
The Atrium, Southern Gate,
Chichester PO19 8SQ, UK

National 01243 779777
International (+44) 1243 779777
e-mail (for orders and customer service enquiries): cs-books@wiley.co.uk
Visit our Home Page on http://www.wileyeurope.com
or http://www.wiley.com

Other Wiley Editorial Offices

John Wiley & Sons Inc., 111 River Street, Hoboken, NJ 07030, USA

Jossey-Bass, 989 Market Street, San Francisco, CA 94103-1741, USA

Wiley-VCH Verlag GmbH, Boschstr. 12, D-69469 Weinheim, Germany

John Wiley & Sons Australia Ltd, 33 Park Road, Milton, Queensland 4064, Australia

John Wiley & Sons (Asia) Pte Ltd, 2 Clementi Loop #02-01, Jin Xing Distripark, Singapore 129809

John Wiley & Sons Canada Ltd, 6045 Freemont Blvd, Mississauga, Ontario, Canada L5R 4J3

Wiley also publishes its books in a variety of electronic formats. Some content that appears in print may not be available in electronic books.

Novartis Foundation Symposium 279
x + 238 pages, 23 figures, 5 tables

British Library Cataloguing in Publication Data

A catalogue record for this book is available from the British Library

ISBN-13 978-0-470-02656-4 (HB)

Typeset in 10½ on 12½ pt Garamond by SNP Best-set Typesetter Ltd., Hong Kong
Printed and bound in Great Britain by T. J. International Ltd, Padstow, Cornwall.
This book is printed on acid-free paper responsibly manufactured from sustainable forestry, in which at least two trees are planted for each one used for paper production.

Contents

Symposium on Innate immunity to pulmonary infection, held at the Wolfson Pavillion, University of Cape Town Medical School, South Africa, 28–30 November 2005

Editors: Derek J. Chadwick (Organizer) and Jamie Goode

This symposium is based on a proposal made by Siamon Gordon and Gordon Brown

Participants

Eric Bateman University of Cape Town Lung Institute, George Street, Mowbray 7700, PO Box 34560, Groote Schuur 7937, Cape Town, South Africa

Linda-Gail Bekker Desmond Tutu HIV Centre, Wernher & Beit Building North, Institute of Infectious Disease & Molecular Medicine, Faculty of Health Sciences, University of Cape Town, Anzio Road, Observatory, 7925 Cape Town, South Africa

Solomon R. Benatar Department of Medicine and Centre for Bioethics, University of Cape Town, J46 Old Groote Schuur Hospital, Observatory 7925, Cape Town, South Africa

Gordon Brown Institute of Infectious Disease & Molecular Medicine, Faculty of Health Sciences, University of Cape Town, Lower Ground Floor, Wernher & Beit Building Sth, Groote Schuur Campus, Observatory, 7925, Cape Town, South Africa

Arnaud Didierlaurent *(Novartis Foundation Bursar)* Kennedy Institute of Rheumatology, Imperial College London, 1 Aspenlea Road, London W6 8LH, UK

Charles Feldman Department of Medicine, University of the Witwatersrand Medical School, 7 York Road, Parktown, 2193, Johannesburg, South Africa

Olivera J. Finn Department of Immunology, E1040 Biomedical Science Tower, University of Pittsburgh School of Medicine, Pittsburgh, PA 15261, USA

Siamon Gordon *(Chair)* Sir William Dunn School of Pathology, University of Oxford, South Parks Road, Oxford OX1 3RE, UK

Eileen Hoal Department of Medical Biochemistry, Stellenbosch University, PO Box 19063, Tygerberg 7505, South Africa

Tracy Hussell Imperial College London, Kennedy Institute of Rheumatology, 1 Aspenlea Road, Hammersmith, London W6 8LH, UK

Bart N. Lambrecht Department of Pulmonary Medicine, Erasmus MC Rotterdam, Dr Molewatersplein 50, 3015 GE Rotterdam, The Netherlands

Jean-Paul Latgé Pasteur Institute, 25 rue du Dr Roux, Paris 75015, France

Stephen Lawn Institute of Infectious Disease & Molecular Medicine, Faculty of Health Sciences, University of Cape Town, Anzio Road, Observatory, Cape Town 7925, South Africa

Alberto Mantovani Fondazione Humanitas per la Ricerca, Istituto Clinico Humanitas, Via Manzoni 56, 20089 Rozzano University of Milan, Milan, Italy

Bongani Mayosi Cardiac Clinic, E25 New Groote Schuur Hospital, Observatory 7925, Cape Town, South Africa

Eamon McGreal Department of Child Health, Wales College of Medicine, University of Cardiff, Heath Park, Cardiff CF14 4XN, UK

Valerie Mizrahi MRC/NHLS/WITS, Molecular Mycobacteriology Research Unit, DST/NRF Centre of Excellence in Biomedical TB Research, National Health Laboratory Service & University of the Witwatersrand, Hospital Street, PO Box 1038, Johannesburg 2000, South Africa

J. Malik Peiris University of Hong Kong, Department of Microbiology, University Pathology Building, Queen Mary Hospital Compound, Pokfulam, Hong Kong, China

Valérie Quesniaux IEM2815 Molecular Immunology and Embryology, Transgenose Institute, CNRS, 3b rue de la Ferollerie, F-45071 Orleans, Cedex 2, France

Luigina Romani Microbiology Section, Dept of Exp Medicine and Biochemical Science, University of Perugia, Via del Giochetto, Perugia 07122, Italy

Bernhard Ryffel IEM 2815, CNRS, Institut Transgenose, 3B rue de la Ferollerie, 45071 Orleans, Cedex 2, France

Barry Schoub National Institute for Communicable Diseases, Private Bag X4, Sandringham, Johannesburg 2131, South Africa

Anthony Segal Centre for Molecular Medicine, Department of Medicine, University College London, 5 University Street, London WC1E 6JJ, UK

John K. Sheehan Cystic Fibrosis Centre, Department of Biochemistry & Biophysics, University of North Carolina at Chapel Hill, Campus Box 7248, 4019a, Thurston Bowles, Chapel Hill, NC 27599, USA

Edith Sim Department of Pharmacology, University of Oxford, South Parks Road, Oxford OX1 3QT, UK

Robert Sim MRC Immunochemistry Unit, Department of Biochemistry, University of Oxford, South Parks Road, Oxford OX1 3QU, UK

David P. Speert Division of Infectious and Immunological Diseases, Department of Pediatrics, Child and Family Research Institute, Room 377, 950 West 28th Ave, Vancouver, British Columbia V5Z 4H4, Canada

Ralph Steinman Rockefeller University, Box 176, Laboratory of Cellular Physiology & Immunology, 1230 York Avenue, New York, NY 10021, USA

Lafras Steyn Institute of Infectious Disease & Molecular Medicine & Department of Clinical Laboratory Sciences, Faculty of Health Sciences, University of Cape Town, Cape Town 7925, South Africa

Paul van Helden Department of Medical Biochemistry, MRC Centre for Molecular and Cellular Biology, Faculty of Health Sciences, University of Stellenbosch, PO Box 19063, Tygerberg, Western Cape 7505, South Africa

Gerhard Walzl Department of Medical Biochemistry, Faculty of Health Sciences, Stellenbosch University, Tygerberg 7505, South Africa

Robert J. Wilkinson Institute of Infectious Disease & Molecular Medicine, Faculty of Health Sciences, University of Cape Town, Cape Town 7925, South Africa

Timothy J. Williams Leukocyte Biology Section, National Heart & Lung Institute, Sir Alexander Fleming Building, Imperial College London, South Kensington Campus, London SW7 2AZ, UK

Chair's introduction

Siamon Gordon

Sir William Dunn School of Pathology, University of Oxford, South Parks Road, Oxford OX1 3RE, UK

To introduce this meeting I would like to list a few topics that we should be thinking about during this meeting, and then at the end we will come back to this list in our final discussion.

Lung infections and to a lesser extent allergies are important diseases in terms of morbidity and mortality. It is not only tuberculosis (TB) that is a problem: this is the second Novartis Foundation symposium in Cape Town, but while TB will be a major focus of this meeting also, we'll be looking at many other important bacteria and viruses.

We have a nice mix here of clinicians and scientists. With input from the different fields represented here, we have a wonderful human 'model' of mucosal immunity. We have experts here who study not only cellular aspects and antibodies, but also the collectins and surfactants, as well as other extracellular factors such as mucus. We want to see what we can learn from the human studies.

Mucosal immunity is an interesting example of interactions between epithelia and haematopoietic cells. Then of course we have something special: we are in the lung, but we also have the interactions with other systemic aspects of host responses. There are not only acute effects (emphasizing the innate aspects), but also long-term sequelae such as the adaptive immune response that follows, and some of the complications such as fibrosis. They are all part of this initial immune response.

The topic of this meeting is of course a worldwide problem, but it is an appropriate problem to be discussed in South Africa. This is a major health problem in this country, but this does provide a laboratory for us to study things that fortunately aren't seen to the same extent elsewhere.

What are some of the issues in terms of host–pathogen interactions? We know that human populations differ. This raises the question of what determines genetic susceptibility of resistance. One breaking topic is whether polymorphisms in, for example, the Toll-like receptors (TLRs) are significant. There is a lot of research in this area, and I've heard of unpublished work on TLR polymorphisms that have a dramatic effect on diseases such as respiratory syncytial virus.

The lower and upper respiratory tracts differ in terms of their commensal organisms. The lower should be essentially sterile, whereas the nose and rest of the upper respiratory tract are not sterile. This is an interesting problem compared with, say, the gut. There is the popular idea of the hygiene hypothesis; many people are wondering whether the increase in allergy and asthma is in some way related to the increasingly clean environment found in some parts of the world.

Tropism is an interesting question: why are certain agents specifically able to infect certain cells? Is this a property of a particular local cell, or is it something that happens in only some cellular environments? Many cells have the same molecules and only in some is there selectivity of infection. This is particularly relevant to some of the major virus infections. We haven't done justice to the adaptive immune response in this programme. If we had, the meeting would have been a lot longer. Nevertheless, there are some fascinating aspects in which the innate response may or may not be able to influence and skew the adaptive response. The Th1, Th2, T_{regs} and antigen-presenting cells (APCs) all have important roles here, both in terms of inducing an immune response and also suppressing one. This may be a unique property of the airway macrophages. This includes the major biological issue of dormancy. How is it that mycobacteria can persist in some quiescent state within cells? How do we study this? These are issues that are important medically, but difficult to address experimentally.

We mustn't forget we are in the lung. The local environment is something we don't pay enough attention to sometimes. This is an organ of gas exchange, so what is the role of oxygenation in this particular site? What about all those particulates that we inhale? The dust diseases have a long history in South Africa because of the mining industry and also asbestos. There are local surfactants in the lung to consider. And then we have the devastating interaction between the lung and smoking. One of the intriguing issues is whether there is sometimes coinfection between a virus and a bacterium which makes one of the two more virulent or pathogenic. This could be an important issue for influenza. Finally, what are the special features of the vascular bed and the lymphatic drainage?

We also have systemic factors that will influence local disease. There is HIV and infection by opportunistic agents. Again, alcohol is a major and neglected problem in this part of the world, as is poor nutrition. Extrapulmonary parasites are also pervasive and may or may not have an impact on diseases within the lung. A further issue is how emerging infections jump species, and move from one individual in a population to another.

From my point of view this is not only an important subject, but also an opportunity to get research done. We have the opportunity to do translational clinical research in South Africa. We should not ignore animal models, but take advantage of the extensive human material available. The lung is an accessible site. We can obtain sputum, bronchoalveolar lavage, aspirates and even pleural effusions. How can we make the best use of this? A general question is how useful is it to monitor blood when this is not the primary site of infection? Are the cells in the blood aware of what is happening in the lung, or are they on their way to the lung? With microarrays it is now much easier to get a signature of what is happening in blood cells in systemic and local diseases.

I don't have to emphasize that we are facing a major threat with emerging infections such as avian flu, and I look forward to hearing about this. I'd like to throw out the provocative idea that both avian flu and SARS may be diseases of innate immunity. There are other less nasty, newly discovered (or to be discovered) emerging viruses. There may be more pathogens out there than we know of.

Although it is not a major theme at this meeting, we'll be touching on the development of vaccines, for example, for influenza. The use of drugs and antibiotics is a highly important issue, but let us not forget what we can do about preventing some of these diseases.

Lung diseases in South Africa: an overview

Eric D. Bateman and Anamika Jithoo*

*Professor of Respiratory Medicine, Director of Department of Critical Care, University of Cape Town and *Research Fellow, University of Cape Town Lung Institute, Cape Town, South Africa*

Abstract. The profile of both infectious and non-infectious lung diseases in South Africa over the past century reflects prevailing sociopolitical and economic forces. The lung, perhaps more than any other organ system is influenced by poverty, occupation and personal habits. These influences are seen in the association between tuberculosis and pneumoconiosis first described in miners, the increasing prevalence of asthma and smoking-related chronic obstructive pulmonary disease, and the current dual epidemics of tuberculosis and infections associated with the human immunodeficiency virus (HIV). The global prediction for developing countries is that by the year 2020 respiratory diseases (including infections) will account for a large majority of deaths and a considerable burden of disability adjusted life years. The country-wide Demographic and Health Surveys of 1998 and 2003 have provided data on symptom prevalence in South Africa. The Lung Health Survey 2002 performed in Cape Town provides disease prevalence and has identified complex interactions between causative factors and disease. Consistent and biologically plausible associations between smoking and susceptibility to tuberculosis and pneumonia in HIV-infected patients have been reported. These findings are relevant both to the planners of public health interventions, and to researchers exploring disease mechanisms and potential remedies.

2006 Innate immunity to pulmonary infection. Wiley, Chichester (Novartis Foundation Symposium 279) p 4–16

The respiratory system, like the skin, serves a unique function at the air-fluid interface between the body and the external environment. Unlike the skin, its function in gas exchange requires it to be delicate and consequently more vulnerable. To compensate for this, it is equipped with a variety of defence mechanisms varying from physicochemical (cough and mucociliary escalator) to immunological. In spite of these it is the target of disease of both infectious and environmental origin which together account for a large proportion of global all-cause morbidity and mortality. In a World Health Organisation (WHO)-commissioned survey of global prevalence of disease in 1990, four diseases of the respiratory system featured amongst the top 10 causes of mortality—lower respiratory tract infections (LRTI) (third place), chronic obstructive pulmonary disease (COPD) (sixth place), tuberculosis (TB) (seventh) and lung cancer in 10th position (Murray & Lopez 1997). Modelling for the future burden of disease, including estimations of the impact of the rising HIV epidemic, the authors predicted that by the year 2020, COPD would have moved to third position, followed by LRTIs, TB, lung cancer and finally HIV-related deaths (other than chest infections). Since some of these are chronic conditions they are and will continue to be leading causes of lost disability-adjusted life years (DALYs). The authors predicted in developed countries the profile of diseases causing loss of DALYs would be lung cancer (in fourth position) and COPD (ninth position). In developing countries like South Africa, the order and importance of respiratory diseases would be similar to those causing mortality described above (Murray & Lopez 1997). In recognition of the importance of respiratory diseases, especially in developing countries (which comprise the majority of the world population), the World Health Assembly of the WHO resolved in May 2000 to make the prevention and control of chronic respiratory disease (CRD) (World Health Organisation 2000) a priority. This has led to the formation of the Global Alliance Against Lung Disease, launched in March 2006 for the purpose of co-ordinating efforts of the WHO, government and non-governmental agencies and initiatives to address these diseases. Moreover, the twin epidemics of TB and HIV infections have become the focus of intense activity and have been accorded the status of global emergencies. South Africa, in spite of its remarkable and unprecedented political transformation has the misfortune of being, if not in the epicentre, then a major victim of this wave of infectious and chronic respiratory pathology. The origins and forces that have created these waves can be traced through the politics and economics of its colonial period into the modern era, and although the interactions of these forces are many and complex, their combined effect presents a profile of disease that is alarming. First was the creation of fertile soil for spread of the white plague, TB, brought from the 'old world' to the vulnerable populations in Africa. The development of a labour market for unskilled and semi-skilled workers through detribalization into a migrant labour force created conditions that favoured transmission (Packard 1989). Next, but related, was the development of mining, which besides its general pollutant effects exposed the workforce to the fibrogenic and carcinogenic effects of silica and asbestos. In spite of the South African mining industry being a world leader in both deep level mining and in the study of dust-related lung disease (including recognition of increased susceptibility of miners with silicosis to TB, the link between silica exposure and systemic sclerosis, and crocidolite asbestos to mesothelioma), there were unacceptable delays in translating these findings into improvements in working conditions. In the case of asbestos, mining of the highly carcinogenic crocidolite variety continued well after its harmful effects were exposed. The full impact of past mining practices, even those previously considered safe, is only now becoming evident (Churchyard & Corbett 2000, teWater Naude et al 2006). One example of this is the very high burden of lung disease including TB amongst retired miners. In 1987 Cowie & van Schalkwyk reported that the prevalence of silicosis among active miners in the Free State goldfields was only 1% (between 0.87 to 1.38%). However studies performed in 1997 (Steen et al 1997) and 1998 (Trapido et al 1998) in retired gold miners who had returned to their homes in Botswana and Lesotho, revealed radiographic, and often advanced stages of silicosis in more than one third of ex-miners and that between one third and one half also had evidence of current or previous pulmonary TB, reflecting the lifelong susceptibility to TB created by silica dust exposure. When one considers that at its height this industry employed more than half a million men, and that the turnover of miners each year was high, it is evident that mining has created a large pool of persons greatly at risk of developing TB and of perpetuating the epidemic (Churchyard et al 2004, teWaterNaude et al 2006).

The next major development that has impacted on the health of the nation and upon respiratory health in particular is the spread of HIV. For the first 80 years of the century, notwithstanding the situation in mines, there was a steady decline in TB notifications in South Africa, suggesting that the National TB Control strategy was beginning to bear fruit (Packard 1989). However, this trend has been reversed and over the past 15–20 years notification rates have risen to record levels, the consequence of the interaction between TB and HIV infections. Amongst miners, according to Corbett et al (2000), notification rates which ranged from 600 to 800 per 100 000 between 1983 and 1991, began to rise in 1992, reaching 3000 per 100 000 by the year 2000. The majority of the increase has been in HIV-infected persons, who now represent more than 70% of patients presenting for treatment of TB. By 2000 the nationwide prevalence of HIV sero-positivity exceeded 20% amongst women attending for ante-natal care, and has continued to rise. By the mid-1990s life expectancy amongst both women and men, which in South Africa as in most African countries had been increasing, began to fall, and has fallen below 50 years in both men and women.

Although silica exposure and HIV infection have had a major impact upon the TB epidemic in South Africa, the pattern of the epidemic in the Western Cape Province has been an enigma. In this province, as there are no mines, the population has negligible exposure to silica, and it was the last to be affected by the HIV epidemic. Yet over the past 30 years TB notification rates have climbed, particularly in low-income communities, and at a time when their income and nutritional status appeared to be improving. Many potential causes for this 'epidemic within an epidemic' have been considered and explored, and there appears to be no single explanation for the phenomenon. Amongst those that have been considered are nutritional factors (particularly vitamin D deficiency, since Cape Town has wet winters with days of overcast skies which might reduce cutaneous conversion of vitamin D to its more active forms) and alcohol abuse (in some communities alcohol abuse is rife, and often associated with poor nutrition). Genetic susceptibility has also been explored, as has TB strain differences, but without convincing results. More promising has been the study of local factors that favour transmission such as housing and social behaviour. Overcrowding has been a feature of life in the affected areas, the effects of which are aggravated by inclement weather which keeps people indoors. A popular pastime is spending evenings in informal taverns where alcohol is consumed, which results in close contact between residents. Finally, there is the potential impact of smoking, both of tobacco and cannabis. In the Lung Health Survey 2002, performed in two suburbs of Cape Town where the notification rate for bacteriologically confirmed TB was 612 per 100 000 persons (Western Cape Tuberculosis Programme 2002), the prevalence of current smoking amongst males was almost 60% and that among women more than 40%, and cannabis use was recorded in 12% of persons (mainly in males) (Jithoo et al 2003). These levels are well above the national averages for current smoking of 42% for males, and 11% for females aged 15 years and older (Steyn et al 2002).

The association between smoking and TB has received increased attention in recent years, with studies confirming a variety of interactions: risk of infection, transition of infection to disease, severity of pulmonary disease, rate of sputum conversion on treatment, risk of relapse, pulmonary impairment after treatment and mortality. For example, Gajalakshmi et al (2003) performed a case-control study of 78 000 men who had died of disease in rural and urban India. Smokers were at a more than fourfold greater risk of death from TB than non-smokers. In those dying from TB the population attributable fraction, i.e. the proportion of the disease occurrence or mortality in the population attributable to the risk factor (smoking), on the assumption that the association is causal, was more than 50%, higher than that contributed by smoking to the rates of lung and upper respiratory tract cancer, and other respiratory diseases. In the Lung Health 2002 survey, 76% of subjects over the age of 15 years had a positive tuberculin skin test (≥10 mm of induration), and the risk of a positive test was significantly higher in smokers than for never smokers (unadjusted OR = 1.99, 95% CI: 1.62 to 2.45) (den Boon et al 2005). A positive dose–response relationship with pack years was also observed, with those smoking more than 15 pack years having the highest risk (adjusted OR = 1.90, 95% CI: 1.28 to 2.81). Although similar findings have been found in restricted communities like a prison in Pakistan (Hussain 2003), nursing home residents in the UK (Nisar et al 1993), Vietnamese immigrants in Australia (Plant 2002) and migrant farm workers in the USA (McCurdy 1997), this is the first study demonstrating this effect in a cross-sectional survey of an entire community. Moreover, the risk was evident even at very low levels of smoking (1 to 5 pack years), raising the possibility that even passive smoking might be harmful. One previous study has examined the effect of passive smoking on TB infection.

Singh et al (2005) reported a significantly greater risk of infection in children aged 5 years or less who were exposed to an adult with TB if there was concurrent exposure to cigarette smoke.

These interactions between smoking and TB highlight the need for stricter tobacco control, particularly in Africa where TB is out of control, and where communities are being targeted by tobacco companies as promising emerging markets for the sale of tobacco products.

Chronic bronchitis and COPD are further examples of diseases which in Africa and developing nations differ from their counterparts in developed countries. In the latter, the principal cause of with both these diseases—the first characterized by chronic cough and persistent sputum production, without evidence of airflow limitation, and the latter being associated with both respiratory symptoms and lung function abnormality, is considered to be cigarette smoke. However, studies in developing countries confirm major contributions from other factors such as occupational exposures in poorly regulated mines and industries (Hnizdo 1990, 1992), environmental including indoor household pollution (Van Hoorn et al 1996, Grobbelaar & Bateman 1991), TB (Churchyard et al 2001) and other infections, and cannabis use (Chan-Yeung et al 2004). For example in the Demographic and Health Survey of 1998, the first national survey of chronic bronchitis in South Africa, the prevalence was lower than that of countries in Europe (Ehrlich et al 2004). However, in the Lung Health 2002 Survey in Cape Town much higher rates were observed. In the former study the population attributable fractions were 10% for past history of TB, 14% for occupational exposures in men and 14% for smoky domestic fuel exposures in women. Although findings in the Cape Town study were similar, the role of cannabis smoking was found to be important. Perhaps because of its illegal status, there are few data on the impact of cannabis on respiratory disease in South Africa, but it is widely used in some communities and requires further study.

Lack of standardisation of the definition and methods for diagnosing asthma and COPD has made it difficult to compare different studies and sources of information (Ehrlich & Jithoo 2006). This deficiency has been corrected by the formulation of international consensus guidelines by the Global Initiative for Asthma (GINA) and the Global Initiative for Chronic Obstructive Lung Disease (GOLD) (Pauwels et al 2001) for asthma and COPD respectively, and the development of standardised methods for use in prevalence surveys for these diseases. The Burden of Obstructive Lung Disease (BOLD) methodology developed by the GOLD initiative (Buist et al 2005) involves use of a standardized validated questionnaire, standardized lung function testing and centralized data collection and statistical analysis, and is being promoted for widespread use in order to improve detection of the disease and provide a basis for focused intervention programmes. Its first use in Africa was in Cape Town, in the population in whom the Lung Health

Survey 2002 had been performed. Preliminary results have provided physiological confirmation that the high rates of symptoms recorded in the Lung Health Survey 2002 reflect a high prevalence of COPD in both men and women (Jithoo et al 2006).

Conclusions

The prevalence of respiratory diseases in South Africa reflects its political and social past, and exposures of large numbers of people to harmful environments both in the workplace and in the community, with the added catastrophe of the spread of the human immunodeficiency virus and the rampant resurgence of TB. Examination of associations between risk factors and different respiratory diseases confirms the major impact of environmental factors in respiratory diseases, and the complex relationships between them. Some of these interactions are depicted in Fig. 1. The notion of innate immunity may not apply in this setting where complex exposures occur early in life and even antenatally, and where no population may be viewed as 'naïve'. These considerations are as important for researchers involved in the study of mechanisms of disease, as they are for those responsible for devising policies and designing services for the promotion of health and the prevention and treatment of lung disease.

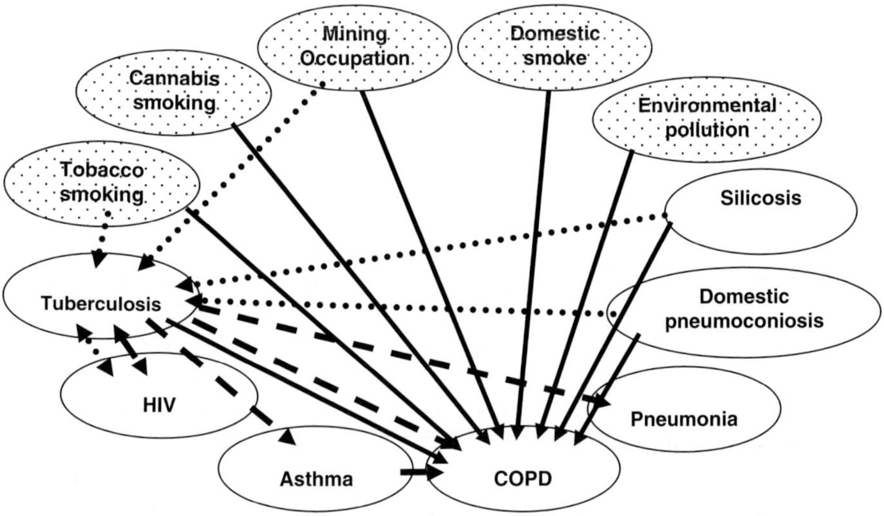

FIG. 1. Recognized interactions between environmental risk factors and common respiratory diseases.

Acknowledgments

Financial support has been obtained from the South African Medical Research Council, Research for International Tobacco Control, and from the South African Thoracic Society through its GlaxoSmithKline Pulmonology Research Fellowship and AstraZeneca Respiratory Research Fellowship for some of the studies in this report.

References

Buist AS, Vollmer WM, Sullivan SD et al 2005 The burden of obstructive lung disease initiative (BOLD): Rationale and design. COPD 2:277–283

Chan-Yeung M, Ait-Khaled N, White N et al 2004 The burden and impact of COPD in Asia and Africa. Int J Tuberc Lung Dis 8:2–14

Churchyard GJ, Corbett L 2000 Tuberculosis and associated diseases 2000 In: Handbook of occupational health practice in the SA mining industry. SIMRAC 2000

Churchyard G, Ehrlich R, te Water Naude JM et al 2004 Silicosis prevalence and exposure-response relationships in South African goldminers. Occup Environ Med 61:811–816

Churchyard GJ, Hnizdo E, White N 2001 Pulmonary tuberculosis in relation to lung function loss. Safety in mines research advisory committee (SIMRAC) Research report: Health 617, SIMRAC, Johannesburg

Corbett EL, Churchyard GJ, Clayton TC et al 2000 HIV infection and silicosis: the impact of two potent risk factors on the incidence of mycobacterial disease in South African miners. AIDS 14:2759–2768

Cowie RL, van Schalkwyk MG 1987 The prevalence of silicosis in Orange Free State gold miners. J Occup Med 29:44–46

den Boon S, van Lill SWP, Borgdorff MW et al 2005 The association between smoking and tuberculosis infection: a population survey in a high tuberculosis incidence area. Thorax 60:555–557

Department of Health 2002 South Africa demographic and Health Survey Report 1998. Department of Health, Pretoria, p 1–338

Ehrlich R, Jithoo A 2006 Chronic respiratory diseases in South Africa. In: Fourie J, Steyn K (eds) Chronic diseases of lifestyle in South Africa. Medical Research Council, Parow, in press

Ehrlich R, White N, Norman R et al 2004 Predictors of chronic bronchitis in South African adults. Int J Tuberc Lung Dis 8:369–376

Gajalakshmi V, Peto R, Kanaka TS, Jha R 2003 Smoking and mortality from tuberculosis and other diseases in India: retrospective study of 43000 adult male deaths and 35000 controls. Lancet 362:507–515

Global Initiative for Asthma (GINA) 1995 Global strategy for asthma management and prevention. NIH Publication 02-3659 Updated 2002 and 2004

Grobbelaar J, Bateman ED 1991 Hut lung-a domestically acquired pneumoconiosis of mixed aetiology in rural women. Thorax 46:334–340

Hnizdo E 1990 Combined effect of silica dust and tobacco smoking on mortality from chronic obstructive lung disease in gold miners. Brit J Ind Med 47:656–664

Hnizdo E 1992 Health risks among white South African goldminers—dust, smoking and chronic obstructive pulmonary disease. S Afr Med J 81:512–517

Hussain H, Akhtar S, Nanan D 2003 Prevalence of and risk factors associated with Mycobacterium tuberculosis infection in prisoners, North West Frontier Province, Pakistan. Int J Epidemiol 32:794–799

Jithoo A, Bateman ED, White NW et al 2003 Prevalence of adult lung disease in a middle-to-low income urban area of South Africa: Lung health survey 2002. (Abstract). S Afr Respir J 9:127

Jithoo A, White NW, Beyers N et al 2005 High prevalence, under-diagnosis and under-treatment of chronic bronchitis in South Africa-an example of differing risk factors in developing countries. (Abstract). Int J Tuberc Lung Dis 9:S283

McCurdy SA, Arretz DS, Bates RO 1997 Tuberculin reactivity among California Hispanic migrant farm workers. Am J Indust Med 32:600–605

Murray CJL, Lopez AD 1997 Alternative projections of mortality and disability by cause 1990–2020: Global burden of disease study. Lancet 349:1498–1504

Nisar M, Williams CS, Ashby D, Davies G 1993 Tuberculin testing in residential homes for the elderly. Thorax 48:1257–1260

Packard RM 1989 White Plague, Black Labor. Tuberculosis and the political economy of health and disease in South Africa. University of Natal Press. Pietermaritzburg

Pauwels RA, Buist AS, Calverley PMA, Jenkins CR, Hurd SS 2001 GOLD Scientific Committee 2001 Global strategy for the diagnosis, management, and prevention of chronic obstructive pulmonary disease. NHLBI Global Initiative for Chronic Obstructive Lung Disease (GOLD) Workshop summary. Am J Respir Crit Care Med 163:1256–1276

Plant AJ, Watkins RE, Gushulak B et al 2002 Predictors of tuberculin reactivity among prospective Vietnamese migrants: the effect of smoking. Epidemiol Infect 128:37–45

Singh M, Mynak ML, Kumar L et al 2005 Prevalence and risk factors for transmission of infection among children in household contact with adults having pulmonary tuberculosis. Arch Dis Child 90:624–628

Steen TW, Gyi KM, White NW et al 1997 Prevalence of occupational lung disease among Botswana men formerly employed in the South African mining industry. Occup Environ Med 54:19–26

Steyn K, Bradshaw D, Norman R et al 2002 Tobacco use in South Africans during 1998: the first Demographic and Health Survey. J Cardiovasc Risk 9:161–170

teWaterNaude JM, Ehrlich RI, Churchyard GJ et al 2006 Tuberculosis and silica exposure in South African gold miners. Occup Environ Med 63:187–192

Trapido AS, Mqoqi NP, Williams BG et al 1998 Prevalence of occupational lung disease in a random sample of former mineworkers, Libode District, Eastern Cape Province, South Africa. Am J Ind Med 34:305–313

Van Hoorn C, Nel R, Terblanche P 1996 Indoor air pollution from coal and wood use in South Africa: an overview. Energy for sustainable development III:38–40

Western Cape Tuberculosis Programme 2002 Health facility report for Uitsig clinic and Ravensmead clinic. City of Cape Town, Cape Town

World Health Organization 2002 Global Burden of disease estimates. [Online]. Available: http://www3.who.int/whosis/menu.cfm?path=burden_estimates [2005, 10 September]

World Health Organisation 2000 Fifty-Third World Health Assembly resolution 53. 17 May

DISCUSSION

van Helden: I've done many simple calculations. From my perspective, the estimate of annual risk of tuberculosis (TB) infection of 3.5% is shocking but also probably an underestimate. While I accept that this figure was calculated using the standard methodology, I don't understand one thing. One sees 25% skin conversion at the age 0–5 to more than 75% at age 15. This means that there is a 50% conversion in 10 years. Purely arithmetically, this is 5% per annum. If you consider that many of these will be dual infections, it must be considerably more than 5%. Can you help me with this?

Bateman: The only correction I would make to your calculation is that the positive tuberculin rate is 75% at age 15 and over: we don't know at which point they convert. The time scale is therefore longer than 10 years. I can't cast additional light on your estimates because the ARI work is not mine. I had hoped that you might have more information on that study. You are correct, though: this calculation was utilizing the standard measure for ARI, which is percentage conversion to positivity in the period of one a year in a susceptible group.

Schoub: I was interested in what you said about the interaction of various factors. Clearly there is a striking relationship between climate and acute respiratory infection. There may be various 'arrows' linking these factors. There may be an epidemiological arrow which you alluded to; there may be a viral triggering factor, and, with relevance to this symposium, there may also be innate immunity factors. I think the climatic factor is one we may want to look into with regard to mechanisms.

Bateman: Climate is one of the most difficult things to study. Keatinge and colleagues have examined the effects of thermal extremes upon mortality. They have demonstrated that people in the colder countries of Europe protect themselves better from cold stress, than those in warmer countries (Keatinge et al 2000). In another study the authors found an inverse association between cold-related mortality figures across six regions of Europe (warmer and colder) and the wearing of gloves, scarves and hats (Donaldson et al 2001). Perhaps what our grandmothers said about dressing warmly, is correct, particularly if you have a weak chest, are frail or elderly. This is indirect evidence and may sound unlikely, but the data are impressive.

Feldman: With pneumococcal infections, there are two studies looking at the influence of ambient temperature. There is a close correlation between the appearance of pneumococcal infections and outside air temperature (Dowell et al 2003). It was thought that one of the risk factors would be viral infections occurring mainly in winter. The question is, why do viral infections occur more commonly in winter? But even if you control for this, and also for the fact that in winter people tend to gather together indoors more, there is still a much higher incidence of pneumococcal infections with cooler air temperature. Meteorological data correlate very closely with this incidence. Even if you factor in HIV, the peak of pneumococcal infections occurs in winter. In HIV-positive individuals the peak still occurs in winter. As a risk factor HIV doesn't overwhelm the effect of ambient temperature on lower respiratory tract infections.

Peiris: The mechanisms of the influence of temperature on respiratory infections is totally unknown. In Hong Kong, a tropical area, the respiratory syncytial virus (RSV) peak occurs in the summer, not the winter. What determines these events? I also wanted to follow up about the role of air pollution and its impact on TB, chronic obstructive pulmonary disease (COPD) and other lung diseases. Can you factor that into these interactions?

Bateman: We collected air quality data for the period of the Lung Health 2002 study. Atmospheric pollution in Cape Town is not heavy, largely because of the strong prevailing winds. It is only on still days that we have high levels of pollution in residential areas. Quantitation of exposures is therefore difficult, and dose–response associations are unreliable. In the Lung Health Study, we did not attempt to relate symptoms to air pollution, so I can't give you any indication of the impact of outdoor pollution. We do have some data on indoor pollution from biomass fuels in rural areas. People who use biomass fuels in poorly ventilated huts have a higher burden of respiratory disease and develop a form of domestic pneumoconiosis. Lesions contain some crystalline quartz, but the major inclusions are carbonaceous. This form of domestic lung disease is however not found in the study area in Cape Town, as electricity is the major energy source for cooking.

Schoub: The observation that RSV is more common in summer in tropical countries is an interesting one. My understanding from studies done in Singapore is that this reflects a climate relationship. Respiratory infection, and in particular influenza, is more common in the rainy season. The common factor may be crowding. This has been well shown in measles where not only incidence but also severity are correlated with the intensity of crowding.

Peiris: You are right. If we take influenza, there are different seasonalities as you go from the temperate regions to the tropics. The reason for this seasonality isn't clear. It isn't a simple matter of temperature and humidity. Nothing seems to correlate.

Speert: I was struggling with how one would handle the massive amount of data and come up with clear correlations, particularly with regard to the association between smoking and TB. The situations where smoking would occur will likely be in pubs, homes and social situations. I don't know how you could tease out these confounding variables to conclude that smoking *per se* contributed to TB.

Bateman: The reviewers for the journals to which we submitted were as cautious as you, and justifiably so. I am not a statistician, but we worked closely with a very good one, and performed multivariate analyses to examine the impact of other factors such as household income (to assess poverty), sex and a variety of other influences, none of which appeared to account for the association. But you are correct; there might have been other unmeasured social influences that contributed to the association. However, the link between smoking and TB infection has been shown in other populations, and we believe that the consistency if not the strength of the association is sufficient to raise concern.

Quesniaux: What about passive smoking?

Bateman: We are currently analysing the children in our study for links between risk of TB infection and parents who smoke. Our initial analysis has demonstrated a positive association, but the signal is weak.

Walzl: I would like to comment on a possible mechanism by which smoking can lead to escape of TB in the context of latent TB infection. There's a recent paper by de Jonge et al (2005), using a mouse model showing that nicotine acts via the a7 subunit of the nicotinic acetylcholine receptor and leads to alternative macrophage activation, with up-regulation of SOCS3 and STAT3.

Lambrecht: It seems that smoking is a risk factor for more severe TB or a higher infection rate. It is also a risk factor for COPD progression. But there are other lung diseases that are protected by smoking, such as extrinsic allergic alveolitis. Garry Anderson from Melbourne claims that smoking leads to somatic mutations in epithelial cells and the stem cell populations in the lung, changing the way that the lung reacts to all kinds of stimuli. Somatic mutations are the first step in cancer and also in all the different immune responses of the lung to the lung diseases. Is there any evidence of somatic mutations influencing the progression of TB?

Bateman: I don't think I can comment other than to say that this is one among many influences that smoking might have in leading to susceptibility to TB infection. Smoke contains a broad range of chemicals with great potential for harm. We need to explore each potential mechanism. These might serve to improve our understanding of the pathogenesis of TB.

E Sim: Measuring the effect of such chemicals on the epithelium is possible: there are techniques available in toxicology. There is methodology available. There are two particular techniques: ^{32}P labelling and very sensitive HPLC and mass spectrometry techniques.

Sheehan: I heard a talk recently on the effect of virus infection on exacerbations on COPD. More generally, are a wider variety of viruses being considered as factors other than just HIV in the development of COPD?

Bateman: It is not easy to do this in an epidemiological setting. Classically it has been done either by direct examination of the tissues or the study of exacerbations and their associations with outbreaks of viral infections. The thinking over the last five years has swung quite strongly towards an infective aetiology both for exacerbations and the progression of COPD.

Sheehan: It is clear that in heavy smokers there is metaplasia and hyperplasia of mucin-secreting cells, with heavy burden of mucus and reliance more on cough for clearance. All of these things suggest that the mucus stasis might be an underlying problem, promoting an environment that is more easily infected. Is this something you might have statistics on?

Bateman: The difficulty is sorting out how smoking results in susceptibility to infection. Together with the hypersecretion of mucus, there is impairment of cilial activity: the mechanisms responsible for moving the mucus blanket.

Sheehan: Stasis of the mucus blanket is accompanied by increased reliance on cough for mucus removal. Data from a cell culture model of cystic fibrosis airway suggests that mucus stasis yields an environment that promotes both bacteria and other infective factors.

Bateman: The point I am making is that mucus is only one of the things that is happening at that time. Other protective mechanisms against viruses are lost. The defences of the mucous membrane are disrupted at many points.

Schoub: There has been some interesting work by Peltola and colleagues a few years ago, looking at the interaction between virus infection and *Streptococcus pneumoniae* in mice. They showed that if you infect mice first with influenza virus, followed by a pneumococcal infection, this aggravates infection (Peltola et al 2005). The same synergistic effect isn't seen if the mice are first infected with the bacterium. The influenza viral infection must in some way damage the innate immunity—perhaps through influenza virus neuraminidase—as suggested by these authors.

Finn: I am interested in the statistics of reinfection with TB rather than reactivation. Those statistics are damning and speak against the hopes of developing a TB vaccine. If even the natural disease cannot generate protection, what do we expect to do with the vaccine? But I wonder whether there is something different about the population that gets easily reinfected versus people who clear the first infection and never get infected. The latter may be the majority of people in whom the vaccine will do just as well as natural disease. This population ought to be studied for genetic predisposition, ability to respond and so on. We should be using the reinfected population as an illustration of how things ought not to be and try harder to generate the vaccine that elicits the type of response elicited in the protected population. In the TB sector, who is really looking closely at reinfection? This is an important issue.

Ryffel: Cannabis is a neuropharmocologically active compound but it may also have neurorespiratory activity. Can you comment on a possible mechanism as to how cannabis acts?

Bateman: I've not been involved in research in this field, but there are recent papers describing the respiratory hazards of cannabis. As far as I am aware the theories are still fairly rudimentary.

Quesniaux: People smoke cannabis, anyway, so you have to add the effect of smoking.

Bateman: Yes, most cannabis smokers also smoke tobacco, and their exposures may be more intense. The favoured local way of smoking cannabis is in a 'hot pipe'; through the neck of a broken bottle into which compacted cannabis mixed with tobacco often methaqualone as well, is packed. This is lit then passed from person to person. The heat of the glass stains the skin in the webspace between thumb and index finger where the bottle neck is held. This level of combustion appears to provide a better 'kick' for the smoker, but also appears to be very damaging to the lungs.

References

de Jonge WJ, van der Zanden EP, The FO et al 2005 Stimulation of the vagus nerve attenuates macrophage activation by activating the Jak2-STAT3 signaling pathway. Nat Immunol 2005 6:844–851. Erratum in: Nat Immunol 2005 6:954

Donaldson GC, Rintamaki H, Nayha S 2000 Outdoor clothing: its relationship to geography, climate, behaviour and cold-related mortality in Europe. Int J Biometeorol 45:45–51

Dowell SF, Whitney CG, Wright C, Rose CE Jr, Schuchat A 2003 Seasonal patterns of invasive pneumococcal disease. Emerg Infect Dis 9:573–579

Keatinge WR, Donaldson GC, Cordioli E et al 2000 Heat related mortality in warm and cold regions of Europe: observational study. BMJ 321:670–673

Peltola VT, Murti KG, McCullers JA 2005 Influenza virus neuraminidase contributes to secondary bacterial pneumonia. J Infect Dis 192:249–257

TB epidemiology and human genetics

Paul D van Helden, Marlo Möller, Chantal Babb, Robin Warren, Gerhard Walzl, Pieter Uys and Eileen Hoal

Department of Medical Biochemistry/(US/MRC) Centre for Molecular and Cellular Biology and DST/NRF Centre of Excellence for Biomedical TB Research, Faculty of Health Sciences, University of Stellenbosch, PO Box 19063, Tygerberg 7505, South Africa

Abstract. The impact of tuberculosis (TB) is considerably lower than one may expect, since in the absence of immunosuppression, fewer than 10% of infected individuals will develop active disease. The relatively low proportion of individuals who progress to active disease after infection can probably be ascribed to innate resistance in most infected individuals, since vaccination using BCG or a previous episode of TB does not work reliably or effectively to confer protection in high burden parts of the world. Innate factors affecting resistance or susceptibility can be modulated by the environment and such external influences cannot be ignored. Specifically, we will address bacterial variability as well as environmental factors such as diet, smoking, helminths and hormones. We will also discuss host genes that may be involved in susceptibility or resistance at various stages of infection or disease. The discovery of as yet unknown genes impacting on TB susceptibility or disease course may lead to new insights into mechanisms of disease and novel therapies. With adaptive immunity being of little value and good TB control programmes being rare, innate resistance is still our best defence against this disease.

2006 Innate immunity to pulmonary infection. Wiley, Chichester (Novartis Foundation Symposium 279) p 17–41

It is estimated that approximately one third of the global population is infected by tuberculosis (TB). Fortunately, far fewer than 10% of those infected eventually become ill (in the absence of immunosuppression). Studies on families, twins and adoptees have suggested that there is clearly a genetic component involved in susceptibility or resistance to TB. On the other hand, recent research has shown that there are many different strains of *Mycobacterium tuberculosis (Mtb)* and that they can influence the course of disease. Clearly we have to deal with on the one hand a successful pathogen, but on the other hand a host that has learnt to live with this pathogen. It has been proposed that the bacterium is perhaps 10 000 years old and evolved as humans domesticated cattle and settled in villages. Recently, however, it has been suggested that the organism may be as old as 3 million years, which implies a long period of co-evolution and thus adaptation on both sides (Gutierrez et al 2005). At least four successive epidemics of tuberculosis are thought to have occurred in the last 4000 years, viz. in the Nile Valley, Greece, the Americas (approximately 1000 years ago) and Europe. Each wave probably spans centuries, but incidence peaks probably last a few decades only.

In general, when *Mtb* is introduced into a naïve population living under harsh conditions, it may spread rapidly. Thus, in the UK with the massive migration into cities that occurred during the industrial revolution and the poor living conditions at that time, an ideal opportunity for an epidemic was created. In London, it was estimated that 20% of all deaths in 1667 were due to TB. TB peaked (possibly) in the UK around 1780 (early industrial revolution) at about 1120/100 000 p.a., or, it is estimated that 1.25% of the entire population died each year from TB. Thereafter, in England, TB started to decline years before other infectious diseases and long before the introduction of control programmes or antibiotics.

A recent mirror of this epidemic was seen in the Inuit, where, after introduction of TB to a naïve settlement, most individuals died and the epidemic rapidly waned. The waning of any epidemic may occur once the living conditions change or the population becomes more resistant, as susceptible individuals disappear. We argue that this is not due to adaptive immunity, but innate immunity which is likely to increase with exposure to the organism and subsequent removal (by death) of susceptible individuals.

Infection by *Mtb* is a complex and multistage process proceeding from the initial encounter with the pathogen. For this reason we need to imagine a multistep process (Fig. 1).

At each stage in this process, innate factors may play an important role. While there is a body of evidence that suggests there may be some immunity acquired from prior exposure to Mycobacterial species (e.g. BCG vaccination), there is also much evidence to suggest that prior infection does not necessarily confer any protection against further infection or progression to disease (Rook et al 2005, Cosma et al 2004).

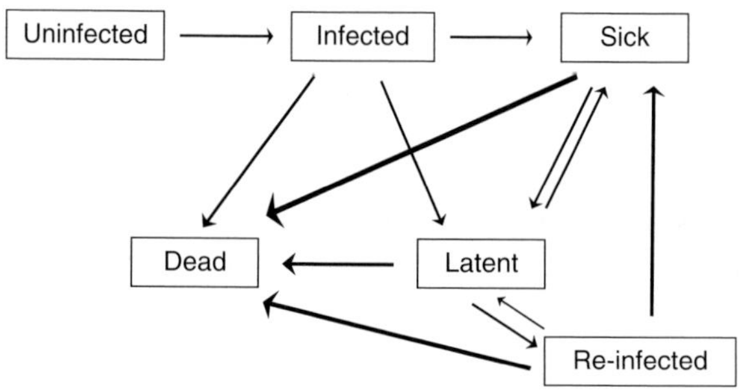

FIG. 1. Transitions in tuberculosis.

Apart from living conditions and an increased (genetic) innate resistance due to death of susceptibles, other factors can play a role: in England and much of the developed world, in the 1800s for the first time food production exceeded population growth and real wages allowed the purchase of adequate food. This is very likely one of the factors that contributed to the decline in TB after 1830, since it is known that a person who is 10% underweight has a threefold increased risk for developing TB after infection.

Excluding sociological determinants, the biological factors likely to be involved in innate resistance to *Mtb* are: (1) the bacterium, (2) the environment (nutrition and other infections), (3) the host. These will be considered below.

The bacterium

Infectious disease studies are complicated by the fact that two genomes, one prokaryotic and one eukaryotic, are interacting in an age-old contest. Genotyping studies have shown that there are thousands of different *Mtb* strains in circulation and comparative genomics has shown that the genome of *Mtb* has evolved through single nucleotide polymorphisms (SNPs), insertions and deletions. This has prompted researchers to investigate the relationship between genome variation and phenotype. A study by Tsolaki et al (2004) suggested a correlation between deletions and the severity of disease while Manca et al (2001) showed that a deletion in the pks 1–15 gene encoding production of phenolic glycolipid was responsible for an altered immune response.

Recent studies have concluded that the 'Beijing' strain is more pathogenic, causes a febrile response on infection and has a higher propensity to develop drug resistance. Furthermore, this strain induces a Th2 immune response on infection allowing for progression towards disease. This differs from the principle genetic group 2 strain CDC1551 which induces a strong Th1 response and less progression towards disease (Manca et al 2001). When the above strains were tested in an *in vivo* mouse model it was shown that the Beijing strain was more pathogenic and could outgrow the CDC1551 strain. There is ample evidence that shows that *Mtb* strains have different growth rates and prompt variable host responses, e.g. cytokine and T cell responses (Manca et al 2001, Janulionis et al 2005, Hoal-van Helden et al 2001a, 2001b). This evidence suggests that these effects are also host dependent. Despite these advances in defining different levels of pathogenicity, many mechanisms underlying these differences in the bacterium remain to be elucidated. It is hypothesised that two evolutionary scenarios may explain these observations: (1) distant evolutionary events which induce an inherited trait that is characteristic of the evolutionary lineage, and (2) recent evolutionary events which induce an inherited trait that is characteristic of a sub-population within a defined lineage.

Thus, signature polymorphisms in *Mtb* isolates may be associated with specific innate (and adaptive) reactions. Some of these also provide a growth advantage and explain the abundance of certain strain types regionally or globally.

The environment

Environmental factors that could be involved in innate resistance to TB include the infection pressure from *Mtb* in the immediate environment, nutrition, other infectious organisms, hormones (e.g. steroid hormones) and substance abuse, such as tobacco smoke (den Boon et al 2005) which suppresses macrophage activity. It is well known that TB can be associated with poverty, which in turn is associated with malnutrition, not only calorie deficit, but more importantly in the case of infectious diseases, with micronutrient imbalances.

Recent work has provided evidence that the different behaviour of omega-3 versus -6 lipids seen at the level of cells and organisms can also be detected *in vitro* in the membrane of the phagosome enclosing mycobacteria. Using phagosomal membrane actin assembly as a functional, *in vitro* readout, these studies showed that the omega-6 lipid, arachidonic acid, as well as six other pro-inflammatory lipids, could stimulate phagosome actin assembly, fusion with lysosomes and a significant increase in pathogen killing. In contrast, the addition of the omega-3 lipids, especially eicosapentanoic acid, suppresses phagosomal actin assembly and induces a significant increase in the growth of pathogenic mycobacteria (*Mtb* and *M. avium*) in macrophages (Anes et al 2003). The ability of these lipids to increase pathogen growth has also been shown to operate at the level of mice and guinea pigs, in the case of both *Mtb* and Salmonella (Paul et al 1997, Chang et al 1992). Our prediction is that dietary manipulation of omega-6 and other pro-inflammatory lipids should help to restrict the growth of pathogens within macrophage phagosomes. Even short term dietary intervention can have dramatic effects on the above-mentioned processes in animal models (Kris-Etherton et al 2002).

The gender bias in tuberculosis has never been satisfactorily explained, but may yield clues to innate resistance of susceptibility factors. Population or gender-based dietary consumption habits would influence disease prevalence. Worldwide, the same gender bias in TB disease is seen. In childhood, no significant differences are noted, but during adolescence girls experience an initially higher rate of TB, whereas in adulthood, males experience a considerably higher disease incidence. Some of this difference in adulthood may be ascribed to behavioural and cultural differences, but other factors are likely to be important. An example of this is the food consumption preferences seen between males and females (see Table 1 below). However, the gender bias may also be linked to steroid hormones, such as DHEA (dehydroepiandrosterone), which have been shown to influence the course of TB

TABLE 1 Gender-based nutritional bias of TB patients

	Mean daily amount	
	Males (n = 23)	*Females (n = 7)*
Fe (mg)	15 (6)	8.7 (8.1)
Mg (mg)	341 (346)	233 (258)
Zn (mg)	14.2 (9.4)	10.1 (6.8)
Se (μg)	71 (45)	30 (45)
Vitamin A (μg)	688 (625)	893 (500)
β-carotene (μg)	2389 (3000–6000)	3522 (3000–6000)
Vitamin B12 (μg)	12.6 (2)	4 (2)
Vitamin C (mg)	67 (75)	108 (60)
Folate (μg)	232 (320)	193 (320)

These figures based on actual food consumed while in hospital (recommended amounts are shown in brackets); see Roberts et al (2005).

disease, specifically at higher concentrations exacerbating pathology (Rook et al 1997).

The net effect of increased iron is to increase risk for active TB, as is the lower levels of some key antioxidant vitamins. The overall effect of these micronutrients and the proteins (such as NRAMP1) involved in their homeostasis (contributing to the 'ionome', Eide et al 2005) is clearly critical. For example, it is known that *Mtb* has an absolute requirement for iron, and that iron supplements should be avoided during TB disease. Furthermore, it has been shown that the total antioxidant profile (could be regarded as a general measure of ion and vitamin status) is significantly lower in TB patients than controls, however, a causative relationship has not yet been established (Wiid et al 2004).

Finally, the effect of multiple infections needs to be considered. Mathematical modelling suggests that in an area of high TB incidence and ARI (annual risk of infection), multiple infection (or super-infection) would be common (Fig. 2).

The simple probability that a particular individual will experience exactly k infection events during a stay of n years (n may be fractional) in a community where the ARI is given by p is:

$$_{\rm n}{\rm P}_{\rm k} = n^k p^k e^{-np} / k! \quad (n = 0, 1, 2, \ldots, k \leq n)$$

Recently, superinfection or reinfection has been proven to occur frequently as predicted (van Rie et al 1999, Warren et al 2004). In an elegant experiment done in a zebrafish model with *M. marinum*, Cosma et al (2004) showed that newly infecting mycobacteria track directly to an existing granuloma harbouring bacteria from a prior infection. The net effect of this is not known, but superinfection may drive

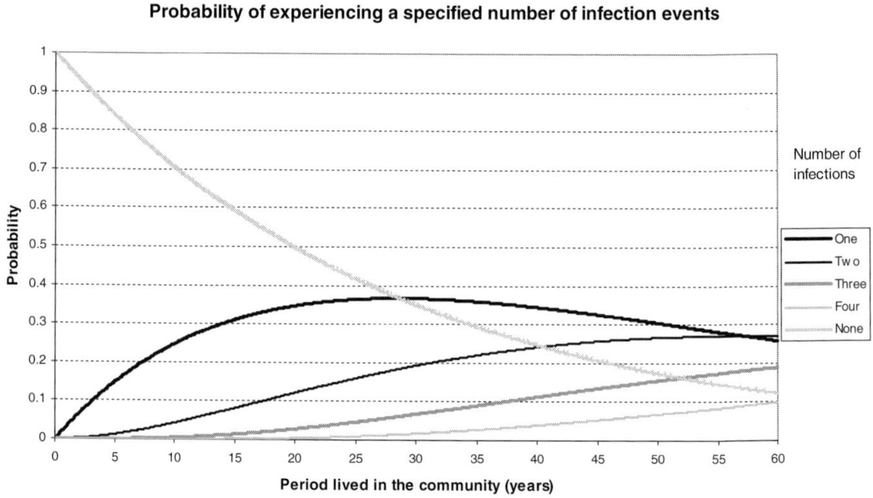

FIG. 2. Muliplicity of Infection. Graphs of ${}_nP_k$ for various k (0, 1 . . . 4), with n along horizontal axis (units of years) for the case of $P = 3.5\%$.

the progression from infection to active disease by either activating latent bacteria (e.g. by means of resuscitation factors, rpf) or by simply overwhelming the innate and adaptive immunity of the host. This effect may be linked to the observation that infection by high doses of mycobacteria will induce a Th2-type response (stimulation of IL4 secretion) rather than Th1, such as may occur on low dose exposure (Rook et al 2005).

Finally, the reaction to infection by *Mtb* is complicated by the intimate connection between the innate and adaptive immune systems and that in reality, most newborn humans are vaccinated with *M. bovis* BCG. Such vaccination may provide for up to 80% protection in developed countries, but far less or none in developing countries, particularly those in the tropics (Rook et al 2005). This may be linked to the mixed Th1/Th2 response in countries of the tropics, which may well be a consequence of exposure to helminths. Thus, the apparent innate response to *Mtb* infection is de facto a 'primed' response and differs in individuals according to exposure to other infecting organisms, such as environmental (myco) bacteria and helminths (Rook et al 2005).

The host

The host defences against intracellular bacteria are mainly cell-mediated but also humoral and therefore any genetic deficiencies in components that play a role in these systems can lead to susceptibility. While there is a definite role for environmental factors, studies have indicated that genetic factors may be even more important than the environment in determining the outcome of infection. The macrophage is usually the first important cell encountering the invading pathogen, and many of the genes governing macrophage function can be expected to influence this essential first step in the innate defence system.

Investigations of TB, a multifactorial disease, have to take into account that there is likely to be an interaction between environmental factors and common polymorphisms in a number of genes. A large body of evidence points to the major role of genetic factors in the human response to a number of infectious pathogens, and these genes could also impact on treatment and vaccine efficacy. The approaches that have been used to identify the genetic component include segregation analysis, animal models and linkage analysis. Understanding the immune responses of individuals with more resistant genotypes, particularly where this can be replicated in a number of different populations, could suggest novel therapies to combat this highly successful pathogen. Complex disease, unlike monogenic conditions, can be influenced by several genes, with each gene making a small contribution to the overall susceptibility to the disease. Tuberculosis is perhaps more complex than most in that the different phenotypes or forms of the disease such as cavitatory TB, pleural effusion, TB meningitis, etc. may be influenced by different genes.

Identification of common TB susceptibility genes

Complex traits such as TB can be investigated via two general designs. Firstly, family-based linkage analysis via genome-wide scanning, and secondly, population-based association studies of candidate genes.

Genome scans

The major advantage of the model-free genome scan is that novel genes may be identified. Although the phenotype is usually TB, it is possible that using intermediate phenotypes in other immune pathways could indicate as yet unsuspected genes. The first genome scan in TB was conducted on two samples of affected sibling pairs from The Gambia and South Africa, and identified two regions, on chromosome 15q and Xq (Bellamy et al 2000). The gene UBE3A in the 15q11–q13 region which encodes a ubiquitin ligase in macrophages, was subsequently associated with TB (Cervino et al 2002). A recent genome-wide scan for tuberculosis and leprosy *per se*, conducted in Brazil, found a cluster of susceptibility genes across chromosome 17q11.2 (Jamieson et al 2004) and indicated that four separate candidate genes, NOS2A, CCL18, CCL4 and STAT5B may contribute to this region of linkage.

Candidate gene association studies

Association studies can suffer from lack of reproducibility of results, and it is important that studies be done with large numbers and repeated, preferably in ethnically diverse populations. However, many of these studies have indicated genes and pathways that are important in the pathogenesis of TB (Fig. 3).

A candidate gene approach: current genes

Human leukocyte antigen

HLA-DR2 is most consistently associated with TB in many populations, including Indian, Polish, Thai, Indonesian and Russian (Lombard et al 2006). Nonetheless, inter-population variations in HLA/TB associations have been reported. HLA-DQB1*0503 was found to influence TB progression in the Cambodian population (Goldfeld et al 1998), but not in the people of the Western Cape (Goldfeld & Hoal, unpublished results). DQB1*0601 was associated with TB susceptibility in the Thai and South Indian population and the HLA haplotype DRB1*08032-DQB1*0601 was associated with genetic susceptibility to multidrug-resistant TB in Korean patients. A study of the Venda population showed an association of DRB1*1302 with TB susceptibility (Lombard et al 2006), whereas Boshoff et al (unpublished data) have shown a marginal association of DRB1*03 with TB in the South African Coloured population.

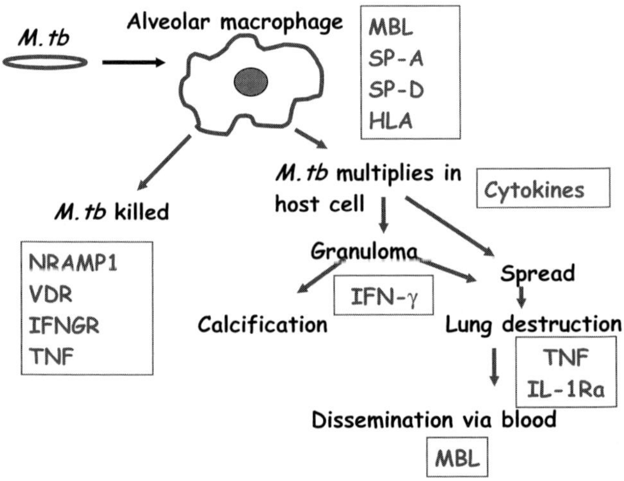

FIG. 3. A simplified representation of the TB disease process and some of the genes that may be involved at different stages.

In this context, it may be worthwhile to explore the relationship between mycobacterial strain type and genetics, e.g. HLA type. For example, in Cambodia, strain diversity is likely to be low and dominated (estimated 60–80%) by the Beijing/W strain type. In the Western Cape, Beijing type strains occur at possibly approximately 20% of total isolates (unpublished data). Therefore, it may be that HLA type is closely linked to *Mtb* strain type in a given locality, and that after extensive exposure, a skewing of HLA type may occur with concomitant resistance to certain strain types. Introduction of new strain types with new epitopes to that locality or ethnic group would then be expected to generate a new epidemic.

Natural resistance-associated macrophage protein

The Natural resistance-associated macrophage protein 1 (*NRAMP1* or *SLC11A1*) gene is a major determinant of natural resistance to intracellular infections, and was originally identified in the mouse model. It is an integral membrane protein expressed only in the lysosome of macrophages and monocytes. After phagocytosis of bacteria, NRAMP1 is targeted to the membrane of the phagosome containing the bacterium, where it may modify the environment to affect the replication of the bacterium, acting as a divalent cation pump which could remove iron or other divalent cations from the phagosome (Blackwell et al 2000).

Associations of *NRAMP1* with TB have been found in Japan, Canada, Korea, Guinea-Conakry, Vietnam, the Gambia and South Africa (Hoal et al 2004) and in most instances the allele over-represented in controls is thought to drive the highest rate of transcription of the protein. Stepwise logistic regression analysis of the South African results indicated that the 5′ and 3′ polymorphisms contribute separate main effects (Hoal et al 2004). More recently, it has been suggested that NRAMP1 may influence only the speed of progression from infection to disease (Malik et al 2005).

Vitamin D receptor

Vitamin D receptor (VDR) is synthesized in monocytes and activated T and B lymphocytes. Its ligand, the active metabolite of vitamin D, calcitriol, is produced in the kidney and by activated monocytes and macrophages, particularly in granulomas. Through its interaction with vitamin D, the retinoid X receptor (RXR) and the vitamin D response element (VDRE), VDR exerts several immunomodulatory effects (Selvaraj et al 2004). These include the activation of monocytes and cell mediated immunity, modulation of the Th1–Th2 host immune response, suppression of lymphocyte proliferation and restriction of *Mtb* survival in macrophages. Vitamin D deficiency is linked to TB by epidemiological evidence. It was found that the prevalence of both vitamin D deficiency and TB was high in Asians because of their vegetarian diet and thus VDR polymorphisms should not necessarily be viewed in isolation. A particular allele was associated with female patients in an Indian population, and was found to increase susceptibility to pulmonary TB in the Gujerati population, but only in patients with a lack of serum vitamin D (Wilkinson et al 2000). The *Fok1* polymorphism was associated with TB in the Chinese Han population. A large study in West Africa found no association in a case control analysis but an association was found with a particular haplotype in the transmission disequilibrium test family data (Bornman et al 2004). This dependence on the haplotype could explain many of the divergent findings on this and other genes.

Evidence of the subtlety of the effect of VDR polymorphisms in the immune response was found by Roth et al (2004) in Peru, who detected an association with time to sputum conversion in TB patients after diagnosis, but did not find a significant association with susceptibility to TB disease.

Collectins

Mannose-binding lectin. Mannose-binding lectin (MBL) is a serum lectin which acts as an opsonin to promote phagocytosis. Intracellular microorganisms may increase their infectivity by using this system, as it promotes the uptake of bacteria into macrophages where they survive. Low functional MBL-serum levels can occur because of the presence of three variant alleles which lead to an unstable protein. Low MBL levels can protect against infection with *Mtb*. This was found in case-control studies where heterozygosity for the *MBL* variant alleles was associated with protection against the disease and the B allele has also been associated with protection against TB and particularly tuberculous meningitis in South Africa (Hoal-van Helden et al 1999). Conversely, an increased susceptibility to pulmonary TB was found in homozygous carriers of the variant alleles in India and a study in Texas gave equivocal results.

Surfactant proteins (SP)-A and SP-D. Uptake of *Mtb* appears to be facilitated by SP-A and inhibited by SP-D. A Mexican population was typed for polymorphisms in both SP-A and SP-D (Floros et al 2000) and TB cases were compared with two control groups. Using multiple logistic regression analysis, an allele of SP-D was found to be associated with susceptibility to TB only when compared with the skin-test positive control group and an allele each of SP-A1 and SP-A2 was associated with TB susceptibility only when compared with the general control group. This illustrates the extreme sensitivity of association studies to definition of phenotype.

Interferon γ/IL12 pathway

Interleukin (IL)12 stimulates interferon (IFN)γ production by lymphocytes, induces type 1 helper T cell responses and is essential for resistance against infection with intracellular bacteria. It is produced by macrophages particularly when infection with intracellular microorganisms occurs. IL12 is a cytokine composed of a heavy chain (IL12B) and a light chain (IL12A). The functional response of lymphocytes to IL12 is dependent on the expression of the IL12 receptor. Any deficiency in these genes will cause a decrease in IFNγ production.

This pathway has been implicated in TB susceptibility by a wide variety of methods. In the mouse model, gene knockout experiments have indicated the importance of IFNγ, IFNγ receptor 1, and IL12 in susceptibility to mycobacterial infection. In this way, we have a gain of knowledge by loss of function. In the human equivalent, specific gene defects have been found to cause rare familial susceptibility to normally non-pathogenic mycobacteria (Casanova & Abel 2002). The first defect identified in a gene in the IFNγ pathway was the autosomal recessive IFNγ receptor ligand binding (IFNγR1) deficiency. This resulted in an overexpression of a dominant form of the IFNγR1, which binds IFNγ, but lacks the intracellular signalling domain. Detailed investigation of patients has led to the identification of mutations in several autosomal genes in the IFNγ pathway that can increase susceptibility to these atypical mycobacterial infections, including complete IFNγ receptor signal transduction chain (IFNγR2) deficiency, autosomal-dominant partial deficiency of the signal transducer and activator of transcription, and autosomal-dominant partial deficiency of the IFNγR1. An autosomal recessive IL12 deficiency was associated with BCG and *S. enteritidis* infection, as were IL12 receptor deficiencies.

At the population level, association with TB was found with a haplotype of *IL12RB1* in Japan, with an intron 2 allele and a specific haplotype in a large study in Hong Kong, and with 2 promoter polymorphisms in a family-based study in Morocco (Remus et al 2004). In a Croatian population an allele of a polymorphic microsatellite of IFNγR1 was associated with protection against pulmonary TB, but not in a Gambian population (Newport et al 2003).

A promoter polymorphism (+874 A→T) in the IFNγ gene itself, which appears to result in lower NF-κB binding and lower transcription levels of IFNγ, was demonstrated in case-control studies to be associated with susceptibility to TB in Sicily, Spain and South Africa (Rossouw et al 2003). The finding in the South African population was replicated in an independent TDT study, confirming the importance of this gene in tuberculosis at the population level.

DC-SIGN

The transmembrane C-type lectin DC-SIGN (Dendritic Cell Specific Intercellular adhesion molecule [ICAM]-Grabbing Nonintegrin), or CD209, is known to be the major *Mtb* receptor on human dendritic cells. DC-SIGN was found to interact with HIV in 1992, and is now also known to be a pathogen receptor for Cytomegalovirus, Ebola, *Helicobacter pylori*, *Leishmania* and *Shistosoma mansoni*. Unifying features of all these pathogens is that they cause chronic infections that can last a lifetime, and their persistence depends on the manipulation of the Th1/Th2 balance. DC-SIGN binds strongly to mycobacteria such as *Mtb* and *M. bovis* BCG via the mannose capped cell wall component (ManLAM) of the pathogen, but does not bind to LAM that lacks the mannose cap (AraLAM). This is intriguing as ManLAM is abundant in slow growing virulent mycobacteria, such as *Mtb* and *M. leprae*, whereas AraLAM is abundant in fast growing atypical, avirulent mycobacteria, such as *M. smegmatis* and *M. chelona*. It has been suggested that *Mtb* targets DC-SIGN both to infect dendritic cells and to down-regulate the dendritic cell mediated immune response.

We tested whether polymorphisms in DC-SIGN are associated with susceptibility to tuberculosis, and found an association between DC-SIGN promoter variation and risk of developing tuberculosis in our South African cohort. The −871G and −336A allelic combination is significantly overrepresented among healthy controls ($P = 1.6 \times 10^{-3}$) and population stratification was excluded (Barreiro et al 2006). The above allelic combination is usually confined to Eurasian populations, and it is possible that these two variants may have increased in frequency in non-African populations as a result of host genetic adaptation to a longer history of exposure to tuberculosis.

Conclusions

Genetic studies in infectious disease are usually complicated because of the presence of two different genomes and the influence their interaction can have on the disease. Although several genes have been identified as susceptibility genes for a number of intracellular bacteria, it is necessary to bear in mind that other genes and the environment can have an influence on the development of the disease, which is the reason that no single major susceptibility gene has been identified in any infectious human disease. The results from strategies used to identify candidate genes or to associate the candidate genes with infectious disease are not the final word on the subject of susceptibility, but provide important evidence on the pathways involved. A greater understanding of the immune response to TB could provide insights into novel treatments that target genetically based susceptibility, such as aerosolised IFNγ, TNF modulation, or even simple supplementation of vitamin D. These therapies could specifically target the more vulnerable individuals in a population and lead to improved health in the entire community. It is likely that each individual has a spectrum of risk factors, which will include genes and environmental factors that will confer a risk profile on that individual. Evidence for this is work that shows that individuals who have had a prior episode have a fourfold higher risk for developing another episode of active TB than those who have never had active TB (Verver et al 2005).

Thus, it may be that innate immunity is the most important process protecting individuals against tuberculosis and by understanding this process; we may develop new ways to combat this ancient scourge.

References

Anes E, Kühnel MP, Bos E, Moniz-Pereira J, Habermann A, Griffiths G 2003 Selected lipids activate phagosome actin assembly and maturation resulting in killing of pathogenic mycobacteria. Nat Cell Biol 5:793–802

Barreiro LB, Neyrolles O, Babb CL et al 2006 Promoter variation in the DC-SIGN-encoding gene CD209 is associated with tuberculosis. PLoS Medicine 3:e20

Bellamy R, Beyers N, McAdam KP et al 2000 Genetic susceptibility to tuberculosis in Africans: a genome-wide scan. Proc Natl Acad Sci USA 97:8005–8009

Blackwell JM, Searle S, Goswami T, Miller EN 2000 Understanding the multiple functions of Nramp1. Microbes Infect 2:317–321

Bornman L, Campbell SJ, Fielding K et al 2004 Vitamin D receptor polymorphisms and susceptibility to tuberculosis in West Africa: 1 case-control and family study. J Infect Dis 190:1631–1641

Casanova JL, Abel L 2002 Genetic dissection of immunity to mycobacteria: The human model. Annu Rev Immunol 20:581–620

Cervino ACL, Lakiss S, Sow O et al 2002 Fine mapping of a putative tuberculosis susceptibility locus on chromosome 15q11-13 in African families. Hum Mol Genet 11:1599–1603

Chang HR, Dulloo AG, Vladoianu IR et al 1992 Fish oil decreases natural resistance of mice to infection with Salmonella typhimurium. Metabolism 41:1–2

Cosma CC, Humbert O, Ramakrishnan L 2004 Superinfecting mycobacteria home to established tuberculous granulomas. Nat Immunol 5:828–835

den Boon S, van Lill SWP, Borgdorff MW et al 2005 Association between smoking and tuberculosis infection: a population survey in a high tuberculosis incidence area. Thorax 60: 555–557

Eide DJ, Clark S, Nair TM et al 2005 Characterization of the yeast ionome: a genome-wide analysis of nutrient mineral and trace element homeostasis in Saccharomyces cerevisiae. Genome Biology 6:R77

Floros J, Lin HM, Garcia A et al 2000 Surfactant protein genetic marker alleles identify a subgroup of tuberculosis in a Mexican population. J Infect Dis 182:1473–1478

Goldfeld AE, Delgado JC, Thim S et al 1998 Association of an HLA-DQ allele with clinical tuberculosis. JAMA 279:226–228

Gutierrez MC, Brisse S, Brosch R et al 2005 Ancient origin and gene mosaicism of the progenitor of mycobacteriumtuberculosis. PLoS Pathog 1:e5

Hoal-van Helden EG, Epstein J, Victor TC et al 1999 Mannose-binding protein B allele confers protection against tuberculous meningitis. Pediatr Res 45:459–464

Hoal-van Helden EG, Hon D, Lewis L-A, Beyers N, van Helden PD 2001a Mycobacterial growth in human macrophages: Variation according to donor, inoculum and bacterial strain. Cell Biol Int 25:77–81

Hoal-van Helden EG, Stanton L-A, van Helden PD 2001b Diversity of *in vitro* cytokine responses by human macrophages to infection by *Mycobacterium tuberculosis* strains. Cell Biol Int 25:83–90

Hoal-van Helden EG, Lewis LA, Jamieson S et al 2004 SLC11A1 (NRAMP1) but not SLC11A2 (NRAMP2) polymorphisms are associated with susceptibility to tuberculosis in a high incidence community in South Africa. Int J Tuberc Lung Dis 8:1464–1471

Jamieson SE, Miller EN, Black GF et al 2004 Evidence for a cluster of genes on chromosome 17q11–q21 controlling susceptibility to tuberculosis and leprosy in Brazilians. Genes Immun 5:46–57

Janulionis E, Sofer C, Schwander SK et al 2005 Survival and replication of clinical mycobacterium tuberculosis isolates in the context of human innate immunity. Infect Immun 2595–2601

Kris-Etherton PM, Harris WS, Appel LJ 2002 Fish consumption, fish oil, omega-3 fatty acids and cardiovascular disease. Circulation 106:2747–2757

Lombard Z, Brune AE, Hoal EG et al 2006 HLA class II disease associations in southern Africa. Tissue Antigens 67:97–110

Malik S, Abel L, Tooker H et al 2005 Alleles of the NRAMP1 gene are risk factors for pediatric tuberculosis disease. PNAS 34:12183–12188

Manca C, Tsenova L, Bergtold A et al 2001 Virulence of a Mycobacterium tuberculosis clinical isolate in mice is determined by failure to induce Th1 type immunity and is associated with induction of IFN-alpha/beta. Proc Natl Acad Sci USA 98:5752–5757

Newport MJ, Awomoyi AA, Blackwell JM 2003 Polymorphism in the interferon-gamma receptor-1 gene and susceptibility to pulmonary tuberculosis in The Gambia. Scand J Immunol 58:383–385

Paul KP, Leichsenring M, Pfisterer M et al 1997 Influence of n-6 and n-3 polyunsaturated fatty acids on the resistance to experimental tuberculosis. Metabolism 46:619–624

Remus N, El Baghdadi J, Fieschi C et al 2004 Association of IL12RB1 polymorphisms with pulmonary tuberculosis in adults in Morocco. J Infect Dis 190:580–587

Roberts T, Herselman M, Marais D, Labadarios D 2005 Served versus actual nutrient intake of hospitalised patients with tuberculosis as compared with energy and nutrient requirements. South African Journal of Clinical Nutrition 18:78–93

Rook GA, Hernandez-Pando R 1997 Pathogenetic role, in human and murine tuberculosis, of changes in the peripheral metabolism of glucocorticoids and antiglucocorticoids. Psychoneuroendocrinology 22 Suppl 1:S109–113

Rook GAW, Dheda K, Zumla A 2005 Immune responses to tuberculosis in developing countries: implications for new vaccines. Nat Rev Immunol 5:661–667

Rossouw M, Nel HJ, Cooke GS, van Helden PD, Hoal EG 2003 Association between tuberculosis and a polymorphic NF-κB binding site in the interferon γ gene. Lancet 361:1871–1872

Roth DE, Soto G, Arenas F et al 2004 Association between vitamin D receptor gene polymorphisms and response to treatment of pulmonary tuberculosis. J Infect Dis 190:920–927

Selvaraj P, Chandra G, Jawahar MS, Vadya Rani M, Nisha Rajeshwari D, Narayanan PR 2004 Regulatory role of vitamin D receptor gene variants of BsmI, ApaI, TaqI, and FokI polymorphisms on macrophage phagocytosis and lymphoproliferative response to mycobacterium tuberculosis antigen in pulmonary tuberculosis. J Clin Immunol 24:523–532

Tsolaki AG, Hirsch AE, DeRiemer K et al 2004 Functional and evolutionary genomics of Mycobacterium tuberculosis: Insights from genomic deletions in 100 strains. Proc Natl Acad Sci USA 101:4865–4870

Van Rie A, Warren RM, Richardson M et al 1999 Exogenous reinfection is a common cause of tuberculosis recurrence after cure. New Eng J Medicine 341:1174–1179
Verver S, Warren RM, Beyers N et al 2005 Rate of reinfection tuberculosis after successful treatment is higher than rate of new tuberculosis. Am J Respir Crit Care Med 171: 1430–1435
Warren RM, Victor TC, Streicher EM et al 2004 Patients with active tuberculosis often have different strains in the same sputum specimen. Am J Respir Crit Care Med 169:610–614
Wiid I, Seaman T, Hoal EG, Benade AJS, Paul D van Helden 2004 Total antioxidant levels are low during active TB and rise with anti-tuberculosis therapy. IUBMB Life 56:101–106
Wilkinson RJ, Llewelyn M, Toossi Z et al 2000 Influence of vitamin D deficiency and vitamin D receptor polymorphisms on tuberculosis among Gujarati Asians in west London: a case-control study. Lancet 355:618–621

DISCUSSION

Segal: What is known about the cellular molecular mechanisms by which the TB bacillus is killed?

van Helden: I don't know where to start in answering this question. For example, the whole notion of nitric oxide (NO) is controversial. Acidification of the phagosome is important, and *Mtb* stops that. Your question to me addresses something I have often said: we shouldn't be looking at TB patients but rather those people who are infected and don't become ill.

Segal: This is the key for understanding the effects of nutrition and IFNγ and so on. We have been doing a study on Crohn's disease, which is similar in that it is often a granulomatous lesion. We did a range of studies in humans rather than in models. We did two things relevant to TB. First, we made skin windows. Gordon Brown had one of these on his arm. You sandpaper a bit of skin off and see what cells come out. You can also apply things onto the skin and see how that can modulate what happens. In Crohn's we found there is a major defect of the accumulation of neutrophils at skin windows. We believe this to be the primary lesion. Second, we injected killed bacteria to see what effect this had on the acute inflammatory response at the site of infection. We used *Escherichia coli*, but you could use TB that had been irradiated. You might find big differences. Many of the immunological effects people are looking at in these diseases are secondary effects to the failure to clear antigen, for example. Then it is assumed that they have some causal relationship to the disease when actually they don't.

van Helden: In TB there are so many different kinds of responses. The first category of infected individuals don't even develop a granuloma and the bacilli are killed. In the next category the granuloma will form, but the bacilli will remain latent for the rest of the person's life. In the third category, the primary focus will form, and there will then be secondary foci and active disease.

Segal: In the first group, I guess one would say that the organism has been killed, digested and removed. In the second, it has been killed but not digested. In the third it hasn't been killed at all. It would be nice to translate that into actual measurements, either in the body or test-tube.

Quesniaux: When patients are treated with anti-TNF antibodies, one of the adverse effects is the occurrence of infections (Mohan et al 2003) and TB is one of the most prominent of these (Keane et al 2001, Mohan et al 2004, Keane 2005). Here we are really acting on the second category: reactivation of previous infections.

van Helden: Perhaps the best evidence for reactivation disease was work done by Troels Lillebaek (Lillebaek et al 2002). The son developed TB 33 years after the father had had a case. It transpired that they both had exactly the same isolate and no one else in that country had the same isolate. There is no way he could have got it from anyone else, indicating that the TB is held inactivated but not dead.

Quesniaux: TB is kept under active control from the immune response.

Brown: I'd like to go back to the point Tony Segal raised about the molecular mechanism of killing. One of the big holes in TB research concerns the receptors that are involved in uptake, and how this influences the resultant response. Many of the proposed receptors don't seem to fit. For example, CR3 is not expressed in alveolar macrophages. And a recent paper shows that in a normal, uninfected individual DC-SIGN is not expressed in macrophages—it is only induced after infection (Tailleux 2005).

Quesniaux: Are there differences in DC-SIGN expression after infection by different strains of TB?

van Helden: We've never looked.

Speert: I agree completely about CR3 not being expressed in alveolar macrophages. My understanding is that cigarette smoking does up-regulate CR3. Could this be a way of tying together some of the things that we have heard? I have a comment about babies infected with BCG. The ones that get seriously ill and die are the ones with profound immunodeficiency. The other group where BCG immunization is contraindicated is chronic granulomatous disease. Reactive oxygen radicals appear to play some role in protection against tuberculosis.

van Helden: We have now found plenty of HIV-positive children with BCGosis.

Mantovani: I was intrigued by the dietary data that you showed. I remember in my medical education that the suggestion was to overfeed patients with TB. The numbers presented here were low. Were these data borne out by larger numbers?

van Helden: The hospital is very small, and there are no other data. This is a major hole in our research. The other problem with TB patients is that they don't feel well, so they don't eat well. If you measure anything in a newly diagnosed TB patient, it may not be trustworthy, because they haven't been eating well because they haven't been feeling well. But in this country, in general, they are also poor, so they may be malnourished from poverty as well. The work done in mice involves much larger numbers and this is well substantiated. In humans nutritional work is poor.

Mantovani: One of the genetic associations you mentioned is with the IL1RA. Is this with infection or disease manifestations? In other infectious disorders, such as *H. pylori* gastritis, the manifestation is not really related to the IL1RA genotype but to the haplotype, and the balance between the pro and anti-inflammatory stimulant molecules in the IL1 system. Are there any data along these lines?

Hoal: There have been a couple of studies on IL1RA (Bellamy et al 1998, Wilkinson et al 1999). As with a number of association studies, they show slightly different things. Some show the haplotype to be associated but some show there is an effect just with the polymorphism.

Wilkinson: We showed that an extended haplotype was associated with higher delayed type hypersensitivity (DTH). The low producing IL1RA allele linked to the high-producing IL1β associated with DTH (Wilkinson et al 1999).

Bekker: We have been doing some work in a small impoverished community with high HIV and TB rates in Cape Town, looking at Mtb strains in patients with and without HIV. We are seeing W Beijing strain much more in young people who are HIV infected. Has anyone looked at the immune competency of people infected with the W Beijing strain?

van Helden: I would predict that the Beijing in your younger age group is a reflection of an emerging strain. It has been relatively recently introduced into our society. Because it is a more able strain it spreads faster. At a Novartis Foundation Symposium in 1997 I proposed a square box scheme, where I hypothesized that you regard your strains as being represented in different parts of this box. They spread well but don't cause much disease, or they cause lots of disease but don't necessarily spread well (van Helden 1998). Beijing is going to be a position in the box. I think it is a very successful emerging strain that is spreading quickly.

Bekker: The counter to that is a set of recent data again from the community we have investigated in the south of Cape Town. In a cross sectional prevalence study in the community where we have sampled randomly, 10% of the community, it seems that people who have HIV and TB co-infection **are** presenting and **are** in the TB clinic and on treatment. People who are sitting quietly with their disease as yet undiagnosed are the older population who are not HIV infected. Immune competent people live with their disease for a long time in this population, before symptoms push them to seek healthcare, and I guess are able to spread their disease. It may be that W Beijing is the predominant strain in this scenario. The immune incompetent (HIV infected) people pick this particular strain up easily; they are the 'canaries' in the population.

van Helden: The question that we have asked is whether Beijing can go into a latency phase. I don't know. This would be interesting to address.

Wilkinson: We have looked at the proportion of strains that are Beijing in isolates from Red Cross Childrens' Hospital in Cape Town. Even within the last few years the proportion of Beijing has tended to increase. It is difficult to know what the denominator is because this is a hospital rather than a community. The second thing is that some of these strains won't grow very well in the laboratory: they appear to have a resuscitation defect. We speculate that if these strains are unable to resuscitate *in vivo*, teleologically speaking they must continue reinfecting people and therefore do best in environments where there is intense transmission.

Mizrahi: What we see depends crucially on what we can culture. Our tools are blunt, with culturability of *Mtb* strains being one of the limiting factors. The standard genotyping tool, which is based on an insertion element, is also relatively crude. However, with the introduction of higher-resolution tools such as DNA microarrays (Tsolaki et al 2004), we are now in a position to start looking at the association between the ability to reactivate and the genotype of the strain. These are studies that need to be done.

Schoub: I want to comment on the interaction between viral infections and TB. We have tried to mine data from the South African National Health Laboratory Service databank to look at seasonality, and correlate it with, for example, reactivation disease. One can speculate on the mechanism but do you have any comment on the actual observation?

van Helden: We don't have any data.

McGreal: What do we actually know at a structural and functional level when we talk about strain difference in *Mtb*? And also at a functional level what impact does strain difference have in terms of the immune response and possible immune evasion strategies. You indicated some associations with different strains, but how much do we know about that?

van Helden: Very little. We are only now developing microarray technology to answer those questions. Part of the problem is choosing the genome to base the microarray on. If you take your standard H37Rv, it has lots of deletions compared to many of our clinical strains. We need to take multiple strains and fully sequence them, and there are now five *Mtb* genomes available. The Broad Institute in Cambridge, MA, may sequence another 10. We should develop a microarray based on everything we can find.

McGreal: When you said that certain strains are prevalent in certain areas and populations, is that a result of local susceptibility? If you were to take a different genetic scheme would you see the same susceptibility?

van Helden: That's something we want to look at.

Mizrahi: It is important to mention that certain studies have revealed that the associations between strain families of *Mtb* and their human host populations are stable (Hirsh et al 2004, Baker et al 2004). This finding speaks to the issue of TB being an ancestral pathogen of humankind. The associations deduced from studies of tuberculosis in immigrant populations imply that particular lineages of *Mtb* may be highly adapted to certain human populations, and less well adapted to others. This notion is supported by the results of a new and larger study published recently by Peter Small's group (Gagneux et al 2006).

van Helden: On the other hand it is to some extent because the immigrant populations stick together. It is a big confound.

Steinman: Can we have a little more discussion about DC-SIGN? First of all, we find it difficult to detect on most human blood leukocytes. Does anyone have experience otherwise and does it change in patients? We find that DC-SIGN is expressed on the small subset of myeloid dendritic cells (DCs) in blood following culture in IL4 (Granelli-Piperno et al 2006). It would be striking if you suddenly saw DC-SIGN increase in TB, since we are all influenced by the papers (Geijtenbeek et al 2003) saying that ligation of DC-SIGN by lipoarabinans from mycobacteria can block the maturation of DCs. The second thing is, DC-SIGN is seen on the monocyte-derived DCs. This is where it was described, and this seems to be due to induction by IL4. If you add IL4 to a monocyte, in a day you will have lots of DC-SIGN. It is not yet clear what monocyte-derived DCs correspond to *in vivo*. The third thing is what happens in the lymphoid organ. We have found that DC-SIGN is abundant on the macrophages in the lymph node medulla. It is not detectable among most DCs in the T cell area in apparently normal lymph nodes. Because of what has been shown *in vitro* with DC-SIGN we should look at it much more assiduously in disease states.

Lambrecht: We have looked at lung DCs from human lavage samples and DC-SIGN is not found on these either.

Gordon: There are other mannose recognition lectins on macrophages.

Brown: The ability of DC-SIGN to modulate the DC function is very likely. We have shown that signalling through another lectin, Dectin-1 can recruit Syk resulting in the induction of IL2 and IL10.

Steinman: That was the message of Geijtenbeek et al (2003). However as mentioned, the sites of DC-SIGN expression *in vivo* are not clear. We are all assuming that it is present on all DCs and exclusively DCs, but neither seems to be the case.

Brown: A lot of these DC receptors are not DC specific. This is another growing theme.

van Helden: How did you look for the expression?

Steinman: By monoclonal antibodies.

Gordon: What about the other genetic factors? You went over them quite quickly as though none of them really mattered, or all of them mattered a little.

van Helden: I don't think anyone has found a major genetic factor.

Mayosi: Do we know the population-attributable risk for any of the genetic factors?

Hoal: It is always small. This is the issue with genetic studies of infectious diseases or other complex diseases. We will find a number of genes that are important but only have a small contribution. I don't know whether you could have the situation where someone's susceptibility rises significantly if they have a combination of a number of these susceptibility alleles. We just don't know that yet. The field is still in the phase of finding the genes, replicating them in a second population, and working out what sort of significance they have.

Gordon: Do you think it is fair to talk about the Bronte family and genetics? How do you separate out the environmental factors and the genetic ones?

van Helden: Apparently, in the Bronte family the father had chronic TB. He died in his 80s, but I think he infected the rest and they all died young.

Gordon: Why does this indicate genetic susceptibility?

van Helden: Other families have a case but not all succumb to disease.

Lambrecht: One clinical problem we run into occasionally is elderly women with atypical mycobacteria in the middle lobe. People always claimed that there were anatomical problems with the middle lobe, but isn't there also a genetic predisposition to develop atypical mycobacterial disease in the lungs?

van Helden: I have no idea. The problem with that sort of thing is power. We looked at our first 2000 isolates, and out of these we found 60 that were attributable to MOTTs. Only now have we put some effort into trying to find this out. But if we only have 60 cases out of 2000, we don't have enough genetic power to really examine this.

Lambrecht: It would be easy to do a candidate gene approach with 60 patients.

van Helden: The populational attributable risk is quite small per gene. It won't be a monogenic effect like the IFNγ receptor defects in the Maltese kindred, for example. You could be right, but I doubt it.

Wilkinson: There was a study done by the British Thoracic Society that looked at susceptibility to *Mycobacterium malmoense* and this showed a weak association with VDR promoter polymorphisms (Gelder et al 2000). However, the predominant factors that predispose people to atypical mycobacterioses are damaged lungs and immunosuppression.

Finn: I know that in this meeting we are focused on the innate immune system. You mentioned several times that this genetic difference may signify that someone's innate immunity is stronger than other people's. One of the important roles of the innate immune system is to jump-start the adaptive immune system and generate good memory responses that will protect you for the rest of your life. The question is, how good is the memory response to the bug in those people who do not get reinfected and those who are protected under high risk conditions? And what types of effector mechanisms are involved that would instruct us in terms of generating that type of immunity through vaccination to protect the rest of the population?

van Helden: This is an important issue, but we haven't done any work on this. There is old work which suggests that the partial protective effect of infection varies between 16 and 18%. Unfortunately, I am not qualified to comment directly on how good that work is. It is based on mathematical modelling and epidemiology. I am sceptical about the interpretation. The tool that is used is generally the skin test, which is extremely blunt. Half of the T cell researchers would say that there is partial immunity. Then there are others who say that it is impossible, because reinfection is occurring and if you are reinfected you have a four times higher likelihood to progress to disease, so where is your partial immunity. To reconcile these two schools of thought I would like to suggest that you increase your categories of individuals. You have categories of individuals that will develop partial immunity and another category that won't.

Finn: Among those that develop partial immunity there will be some that will still be susceptible.

Brown: Is it going to be possible to make vaccine for TB, given the occurrence of reinfection?

van Helden: Many think so, since it could be cell-mediated immunity.

Finn: With regard to the comment you made about a low dose of BCG maintaining better protection, to an immunologist this immediately says that it generates higher affinity and avidity T cells and so on.

van Helden: The New Zealanders say that the dose and timing of repeat BCG is critical. The Irish say that a high-dose BCG vaccine given once is just as good.

Walzl: I can't think of any pathogen where a vaccine provides better protection than the natural infection. But there is always a first time.

Steinman: Cowpox isn't natural for humans, but it worked as a smallpox vaccine. Immunologists just haven't got onto the vaccine scene, but when we do, I think that we can do better than complex microbial vaccines!

Finn: As you mentioned, some pathogens have evolved a symbiosis with the host, and a vaccine might do a better job eliciting immunity than the bugs.

Ryffel: Going back to Valerie's comment on the anti-TNF antibody induced reactivation of chronic/latent infection, we are able to model reactivation or tuberculosis infection in mice (Botha & Ryffel 2003). By administration of neutralizing TNF antibodies or soluble TNF-R we are able to reactivate chronic infection (unpublished). Further we demonstrated that membrane TNF provides a partial protection to infection (Fremond et al 2005) suggesting that neutralising exclusively soluble but not membrane TNF may reduce the risk of reactivation of TB infection.

Quesniaux: An individual who has had a first infection with TB and has had this under control for 20 years will have mounted a very efficient T cell memory response. If you now come along with anti-TNF, this could all go wrong with the TB being reactivated.

Finn: I disagree with your statement that this person has mounted a good memory response. This is an effector memory-type response, something that is continuously protective. It is not that good deep central memory that results from eliminating the pathogen. In the complete absence of antigen you deeply bury your central memory response that can then be reactivated on subsequent infection. You cannot say that reactivation of the bug by TNF means that it has now defeated a well established memory response.

Quesniaux: How do we explain the effect of anti-TNF?

Steinman: TNF has many effects on DCs. In the literature on NOD mice, people have tried to manipulate the onset of that spontaneous autoimmune disease by manipulating TNF levels. There is now evidence that the DC is responsible. The approach is to block TNF early in life, which then reduces the severity of the disease. If you then target antigens to the DCs in the TNF blocked mice, you can see changes in antigen presentation. Therefore, the DCs may be changing as a result of TNF manipulation.

Finn: There is constant cross-talk between the innate and the adaptive immune system. It doesn't just go from the innate to the adaptive and then stop, it goes back as well. You have a wonderfully stimulated memory CD4 response specific for the pathogen. You can arm your macrophage that is infected with a pathogen to make more IFNγ by the interaction with the $CD4^+$ T cell. If you have a good adaptive immune memory, it can then stimulate much higher activity of your innate system when the new infection comes. If you have a very good $CD4^+$ T cell memory response, that macrophage will be much more effective in destroying the pathogen.

van Helden: In thinking about vaccination and protective immunity, we have to remember that there is evidence that BCG does work. My feeling is that if there is no immunosuppression, 90–95% of people infected with TB will not become ill. BCG is not going to do anything for them anyway. It is a small part of the population in which vaccination can have an effect.

Finn: I would like to argue against this point. The 95% of the people will not develop disease unless they are immunosuppressed in some way, unless they live a long life and their immune systems become old. But if you generate a strong immune memory early in life that immune memory is a bit more protective from what happens to that person later in life, who therefore, if healthy and well-nourished will not be susceptible to the disease. Generating a strong and effective immune memory early in life will protect many more people than the 5% that you think will be protected by the vaccine.

Steinman: The point is how you assay memory. This is an evolving field. We used to say that HIV-infected people have good memory, because you could easily detect $CD8^+$ T cells in them that make IFNγ in response to HIV antigens. By current criteria, though, these CD8 cells are dysfunctional. They don't grow in response to antigen. From the same individuals, CMV, EBV and influenza specific cells grow fine (Arrode et al 2005). There is a dysfunction here of the HIV specific $CD8^+$ T cells, and they are termed 'helpless' memory cells. How we assay memory is very important.

Finn: The bottom line is that we know a lot more than last year and 10 years ago. What we know speaks much more in favour of the vaccine being able to do this better than the natural disease.

Steinman: Then there are suppressor cells. What you described in terms of reinfection would fit what we know about suppressor cells. These cells are triggered by antigen, but then they can suppress other immune responses to other antigens, particularly when they are presented by the same presenting cell. If you respond to mycobacteria A and make a suppressor cell specific for peptides from protein A, it will block the immune response to other mycobacteria proteins, B. That is, as long as the antigen presenting cells are presenting both A and B, the suppressor cells for A will block presentation of B. However, it is still not straightforward to measure suppressor cells currently in human. This is a big gap.

Hoal: I want to return to the vaccination and susceptibility story. This is mouse work that I don't know very well, but I heard that some mouse strains that are more susceptible to mycobacteria are also less likely to be protected by vaccination. This is quite worrying if the same were to occur in the human population. We have to be careful when we do vaccination trials in humans to look at this susceptible portion of the population.

Gordon: Bernard Ryffel, do the mouse models allow you to get at some of these questions?

Ryffel: It is well established that the susceptibility to TB infection differs among mouse strains, B6 mice are typically resistant, while Balb/c mice are more susceptible; the genomic analysis will certainly provide polygenic resistance loci, which are currently unknown. What is really amazing is that mice with complete ablation of Toll-like receptor/MyD88 signalling are able to have preserved T cell response to TB antigens. But this T cell response is not protective, as the innate immune response is profoundly defective (Fremond et al 2004). However, BCG vaccination of MyD88 deficient mice provides a short-term, but not long-term protection.

Walzl: There are mouse models concerning regulatory cells. Strangely enough, people examined induction of regulatory cells by *Mtb* in conjunction with the hygiene hypothesis and the allergic response. There are several studies showing *Mtb* in circulatory cells that suppress allergic airway responses. It is strange that people look at it that way; the logical conclusion would be that those responses are there because *Mtb* wants to promote itself.

Lambrecht: This could explain why the low dose BCG works better than the high dose. If you give high doses of freeze-inactivated BCG it is a strong inducer of T_{regs}. There are moieties within the BCG which have the ability to induce

T_{regs}. If they are to the advantage of the bug, this would explain why high levels of BCG would lead to greater susceptibility. It is a balance between protecting from an over-zealous response and on the other side inducing immunity.

E Sim: Am I right in thinking that you can distinguish between reactivation and reinfection by the same strain?

van Helden: No, we can't.

Ryffel: Do you have more evidence that there is coinfection? If this is the case, with two strains, what is the prevalence? What is the contribution of each?

van Helden: We have no quantitative data.

Wilkinson: There is an interesting recent paper showing three cases of TB, all of which are pulmonary, and have another extrapulmonary site (Garcia de Viedma et al 2005). For each of the three extrapulmonary sites there is a different strain. The authors suggest that the extrapulmonary strain has greater virulence than the pulmonary strains when they infect macrophages together.

References

Arrode G, Finke JS, Zebroski H, Siegal FP, Steinman RM 2005 CD8+ T cells from most HIV-1 infected patients, even when challenged with mature dendritic cells, lack functional recall memory to HIV gag but not other viruses. Eur J Immunol 35:159–170

Baker L, Brown T, Maiden MC, Drobniewski F 2004 Silent nucleotide polymorphisms and a phylogeny for Mycobacterium tuberculosis. Emerg Infect Dis 10:1568–1577

Bellamy R, Ruwende C, Corrah T, McAdam KP, Whittle HC, Hill AV 1998 Assessment of the interleukin 1 gene cluster and other candidate gene polymorphisms in host susceptibility to tuberculosis. Tuber Lung Dis 79:83–89

Botha T, Ryffel B 2003 Reactivation of latent tuberculosis infection in TNF-deficient mice. J Immunol 171:3110–3118

Fremond CM, Yeremeev V, Nicolle DM, Jacobs M, Quesniaux VF, Ryffel B 2004 Fatal Mycobacterium tuberculosis infection despite adaptive immune response in the absence of MyD88. J Clin Invest 114:1790–1799

Fremond C, Allie N, Dambuza I et al 2005 Membrane TNF confers protection to acute mycobacterial infection. Respir Res 6:136

Gagneux S, DeRiemer K, Van T, Kato-Maeda M et al 2006 Variable host-pathogen compatibility in Mycobacterium tuberculosis. Proc Natl Acad Sci USA 103:2869–2873

Garcia de Viedma D, Lorenzo G, Cardona PJ et al 2005 Association between the infectivity of Mycobacterium tuberculosis strains and their efficiency for extrarespiratory infection. J Infect Dis 192:2059–2065

Geijtenbeek TB, Van Vliet SJ, Koppel EA et al 2003 Mycobacteria target DC-SIGN to suppress dendritic cell function. J Exp Med 197:7–17

Gelder CM, Hart KW, Williams OM et al 2000 Vitamin D receptor gene polymorphisms and susceptibility to Mycobacterium malmoense pulmonary disease. J Infect Dis 181:2099–2102

Granelli-Piperno A, Shimeliovich I, Pack M, Trumpfheller C, Steinman RM 2006 HIV-1 selectively infects a subset of nonmaturing BDCA1-positive dendritic cells in human blood. J Immunol 176:991–998

Hirsh AE, Tsolaki AG, DeRiemer K, Feldman MW, Small PM 2004 Stable association between strains of Mycobacterium tuberculosis and their human host populations. Proc Natl Acad Sci USA 101:4871–4876

Keane J 2005 TNF-blocking agents and tuberculosis: new drugs illuminate an old topic. Rheumatology (Oxford) 44:714–720

Keane J, Gershon S, Wise RP et al 2001 Tuberculosis associated with infliximab, a tumor necrosis factor alpha-neutralizing agent. N Engl J Med 345:1098–1104

Lillebaek T, Dirksen A, Baess I, Strunge B, Thomsen VO, Andersen AB 2002 Molecular evidence of endogenous reactivation of Mycobacterium tuberculosis after 33 years of latent infection. J Infect Dis 185:401–404

Mohan AK, Cote TR, Siegel JN, Braun MM 2003 Infectious complications of biologic treatments of rheumatoid arthritis. Curr Opin Rheumatol 15:179–184

Mohan AK, Cote TR, Block JA, Manadan AM, Siegel JN, Braun MM 2004 Tuberculosis following the use of etanercept, a tumor necrosis factor inhibitor. Clin Infect Dis 39:295–299

Tailleux L, Pham-Thi N, Bergeron-Lafaurie A et al 2005 DC-SIGN induction in alveolar macrophages defines privileged target host cells for mycobacteria in patients with tuberculosis. PLoS Med 2:e381

Tsolaki AG, Hirsh AE, DeRiemer K et al 2004 Functional and evolutionary genomics of Mycobacterium tuberculosis: insights from genomic deletions in 100 strains. Proc Natl Acad Sci USA 101:4721–4722

van Helden PD 1998 Bacterial genetics and strain variation. In: Genetics and tuberculosis. Wiley, Chichester (Novartis Found Symp 217) p 178–194

Wilkinson RJ, Patel P, Llewelyn M et al 1999 Influence of polymorphism in the genes for the Interleukin 1 Receptor Antagonist and Interleukin IL-1β on tuberculosis. J Exp Med 189:1863–1874

Bacterial infections of the lung in normal and immunodeficient patients

David P. Speert

Room 377, Child and Family Research Institute, 950 West 28th Avenue, Vancouver BC, Canada, V5Z 4H4

Abstract. The lung is exposed to enormous quantities of air and to potentially infectious agents, but serious infections rarely occur, a testament to the extraordinary natural defences of the respiratory tract. The most common causes of bacterial lung infections in normal hosts include *Streptococcus pneumoniae*, *Haemophilus* species, *Staphylococcus aureus* and *Mycobacterium tuberculosis*. In compromised hosts, the bacterial causes of pneumonia are much broader, including species not usually considered of high virulence in humans. Indeed infection with one of these unusual bacterial species demands a search for an immunocompromising condition. Normal defences of the respiratory tract include non-specific physical factors (the 'mucociliary escalator'), and innate factors, including defensins, lysozyme and phagocytic cells (polymorphonuclear leukocytes, pulmonary alveolar macrophages and dendritic cells). Antibacterial defences are enhanced by opsonins, including those intrinsically present (surfactant and complement components) and induced immunoglobulins. Immunocompromising conditions, in which bacterial lung infections frequently occur, include (but are not limited to) hypogammaglobulinaemia, chronic granulomatous disease and primary ciliary dyskinesia. Each of these conditions illustrates the essential role of the disabled element of the innate and adaptive immune system in maintaining sterility of the lower respiratory tract.

2006 Innate immunity to pulmonary infection. Wiley, Chichester (Novartis Foundation Symposium 279) p 42–55

The endobronchial tree presents an enormous surface that is repeatedly exposed to airborne contaminants. It is not surprising therefore that pneumonia is one of the more common infectious diseases in humans, with two to three million community acquired cases per year in the USA (Segreti et al 2005). The upper and lower airways are protected by many non-specific, innate and adaptive defences against infection (Happel et al 2004); although these are highly effective under most conditions, they can be overwhelmed with the resulting onset of pneumonia. Furthermore, some of these natural defences are defective in specific primary or acquired immunodeficiencies, resulting in recurrent sinopulmonary infections. In this brief review, I will describe the normal defences of the respiratory tract and will illustrate their importance by the infections that develop when specific components are compromised.

Pathogenesis of pneumonia

The lower airway is normally sterile, a state maintained by the actions of host defensive elements described below. However, the many mechanisms which usually maintain lower airway sterility can fail when exposed to a particularly virulent microorganism or because of primary or acquired immunodeficiency. The upper airway is colonized by a wide array of bacteria; this resident microflora appears to prevent other more virulent microbes from establishing colonization (Happel 2004). However, potentially virulent bacteria can be part of the normal flora or can replace the normal flora if it is perturbed, as by antimicrobial therapy. Pneumonia is established when bacteria gain access to the lower respiratory tract after aspiration from the upper airway or inhalation of airborne bacteria (Happel et al 2004). The former appears to occur when mucociliary clearance is perturbed and the latter in diseases such as tuberculosis, in which the causative agent *Mycobacterium tuberculosis* does not colonize the normal upper airway.

Bacteria which gain access to the lower airway cannot necessarily cause pneumonia. For such infection to be established, the microbes must resist phagocytic killing by resident and/or recruited phagocytic cells (Masten 2004, Twigg 2004, Wang et al 2004). Indeed, some bacteria, such as *M. tuberculosis* cause infection by their capacity to survive within phagocytic cells (Mason & Ali 2004). Pneumonia can be acute and self-limited if innate and adaptive immune effectors can control it. Under such conditions, the bacteria first replicate in the lower airway and are then ingested and killed, predominantly by recruited polymorphonuclear leukocytes (PMNs); this is the case in pneumococcal pneumonia. Chronic infection can be established if bacteria gain access to long-lived cells such as macrophages or dendritic cells and resist their bactericidal effects; this is the case in tuberculosis.

Pneumonia usually resolves, often aided by antimicrobial therapy. Resolution may be accompanied by development of acquired and specific humoral and/or cell-mediated immunity. Such immunity then prevents subsequent infection with the same infectious agent.

Host defences of the respiratory tract

Non-specific defence

Sterility of the lower airway is maintained in large part by numerous interrelated components, predominantly in the upper airway. For inhaled bacteria to gain access to the lower airway, they must resist the antimicrobial activities of soluble mediators as well as the cleansing effects of the mucociliary escalator (Happel et al 2004).

Numerous natural antibacterial agents are present in the upper airway, the effects of which potential pathogens must resist. Among these agents are lysozyme, peroxidase and lactoferrin in the saliva. Small cationic peptides (defensins) are both directly antibacterial and immunomodulatory (Bowdish et al 2005). Whether they are present in concentrations sufficient to kill inhaled bacteria has not been determined. However, their many immunomodulatory activities may protect the lung against infection in some circumstances.

The airway is endowed with a network of physical factors which propel inhaled agents to the mouth where they can then be swallowed or expectorated. Hairs protrude from the nasal epithelium that function to screen inhaled bacteria from the air and prevent their progression to the oropharynx. Cilia, hair-like projections from the upper respiratory tract epithelial cells, beat in an organized fashion to propel particles toward the oropharynx. The ciliary epithelium is coated with a layer of mucus that traps inhaled particles and aids in the propulsion of potentially infectious agents toward the mouth. Dysfunction of these principal elements of the mucociliary escalator results in recurrent pneumonia, as described below.

Innate defences

Until recent discoveries about specificity, elements of the innate immune system were described as non-specific. Elements of the innate immune system are largely phagocytic, consisting of cells that are in direct contact with the apical surface of airway epithelia. These cells include PMNs (Wang et al 2004), pulmonary alveolar macrophages (Twigg 2004) and dendritic cells (McWilliam et al 1994). These cells recognize foreign prey and capture them by innate or nonopsonic receptors, which will be discussed in other chapters. The most important class of innate receptors appear to be those of the Toll-like group, of which 10, each with unique specificity, have been described in humans. Defects in Toll-like receptors (TLRs) *per se* have not been described in human, but polymorphisms appear to alter susceptibility to infectious diseases (Cook & Pisetsky 2004, Hawn et al 2005), particularly Legionnaire's disease (Hawn et al 2003, 2005).

PMNs are the principal bactericidal phagocytes of the lower respiratory tract, but are not present until inflammation is induced by foreign challenge, such as infection. PMNs are capable of killing by oxidative (generation of reactive oxygen species [ROS]) and nonoxidative means. Bacteria that are resistant to nonoxidative killing are particularly virulent in the lung when ROS are either not produced (as is the case in chronic granulomatous disease) (Speert et al 1994) or if the redox balance in the lung is perturbed (as occurs in cystic fibrosis).

Adaptive defences

Phagocytic killing is greatly enhanced by opsonins, substances which may or may not be present innately in the lower respiratory tract. Opsonins enhance ingestion via opsonic receptors and may even enhance intracellular killing (a function which has been attributed to serum immunoglobulin and complement (Leijh PC et al 1984). Secretory immunoglobulin (Ig)A, some complement components, collectins and surfactant proteins may all be present in the lower respiratory tract prior to the onset of inflammation. Each of these can function as an opsonin, but none is as effective as IgG which is not present under resting conditions. However, IgG is induced by bacterial challenge or immunization, a fact supported by the substantial reduction in pneumococcal disease after immunization with either of the licensed pneumococcal vaccines (Dear et al 2003, Lucero et al 2004).

Pneumonia in immunocompetent hosts

Most cases of pneumonia in immunocompetent hosts are probably non-bacterial. However, bacterial pneumonia occurs with enhanced frequency in the very young and the elderly. Such infections are either community-acquired or hospital acquired. The latter occur in individuals who often have inherited or acquired immunocompromising conditions, and will be discussed here and in the following section.

Community acquired bacterial pneumonia is usually caused by *Streptococcus pneumoniae*, *Haemophilus influenzae*, *Staphylococcus aureus*, *M. tuberculosis*, *Mycoplasma pneumoniae* and *Chlamydia* species. *M. tuberculosis* is a common cause of pneumonia in the developing world and is one of the leading causes of death worldwide; however it is an uncommon cause of pneumonia in the developed world, probably because of better hygiene and effective strategies for anti-tuberculous therapy. Immunization in the developed world against *H. influenzae* type b (Watt et al 2003) and *S. pneumoniae* (Lucero et al 2004) has virtually eliminated the former and has decreased the prevalence of the latter. However, infections with these agents are still prevalent in the developing world, and even after pneumococcal immunization, are caused by serotypes not contained in the vaccine. Bacterial pneumonia is often (but not always) lobar, but the aetiology usually cannot be established, as culture of the throat does not reveal the pathogen. The only ways to establish the aetiology is by blood culture (which may be positive in up to 25% of cases of pneumococcal pneumonia), by paracentesis of pleural effusions or by transcutaneous lung puncture. The latter procedure is rarely used, having been replaced by bronchoscopy or open lung biopsy. These latter procedures are reserved for the most serious cases and are rarely performed.

Hospital-acquired pneumonia can be caused by the same agents as those from the community, but the list of other pathogens is very broad (Lynch 2001).

Hospitalized patients are exposed to a wide range of potentially infectious agents from other patients, and their capacity to resist infection with agents of lower intrinsic virulence is often diminished. For instance, respiratory pathogens may be spread by droplet or aerosol if proper infection control practices are not followed.

Patients who are mechanically ventilated via an endotracheal tube are at greatly enhanced risk for bacterial pneumonia (Shaw 2005). Such ventilator-associated pneumonia can be caused by a wide range of pathogens which reach the lower airway by aspiration, as the normal functions of the mucociliary escalator are abrogated by the endotracheal tube. In these patients, possible causative agents of pneumonia can be found in the throat, but a definitive aetiology can only be established by the same processes outlined above. The oral flora of individuals on ventilators is perturbed by antibiotic therapy and other poorly understood mechanisms. Therefore Gram-negative bacteria such as *Pseudomonas aeruginosa* are among the leading cause of ventilator-associated pneumonia, but they rarely cause lung infections in normal hosts.

Pneumonia in compromised hosts

Primary ciliary dyskinesia (PCD)

Respiratory epithelial cilia are critical in maintaining sterility of the lower respiratory tract. Therefore infections occur when their function is disordered; such is the case in ciliary dyskinesia (Cowan et al 2001), a condition in which there is an abnormal structure of dynein arms in the cilia rendering them uncoordinated in their propulsive activity. This condition is rare and may be a part of Kartagener's syndrome in which there is situs inversus (reversal of the position of major thoracic and abdominal organs). Patients with ciliary dykinesia are at risk for infection from organisms aspirated from the upper respiratory tract. This and other disorders which compromise mucociliary and cough clearance (e.g. alcoholism) set the stage for infections with bacteria which are of lower virulence than in those individuals with intact function.

Cystic fibrosis (CF)

CF is an autosomal recessively inherited disorder of chloride transport across epithelial cells, affecting about 1 in 2000 Caucasian births. Chronic pneumonia is a prototypical manifestation of CF and is the most common cause of death (Currie et al 2003). By definition, host defence of the respiratory tract is compromised, but a universally-accepted explanation for chronic endobronchial inflammation and infection has not been posited. Infections typically begin during the first decade of life and herald a chronic course of pulmonary decompensation. Infection is initially confined to the endobronchial space, but in the later stages of the disease, bronchiectasis occurs. Two classes of bacteria, *Pseudomonas aeruginosa* (Currie et al 2003) and *Burkholderia cepacia* complex (Speert 2002), are the most common agents of pneumonia in CF, and in many patients chronic infection with the same strain persists from acquisition until death. The strains of *P. aeruginosa* infecting patients with CF are atypical—mucoid, lipopolysaccharide (LPS) rough and non-motile (Currie et al 2003). It appears that transition to this typical 'CF bacterial phenotype' occurs within the endobronchial tree, as the initial infecting strains are usually like those found in nature—non-mucoid, LPS smooth and motile. There is considerable debate about why such a narrow range of bacteria are so commonly seen in CF but rarely in other childhood infections.

No classical primary defect in innate or adaptive immunity has been found in patients with CF to explain their susceptibility to chronic endobronchial infection. It is most likely that there exists a defect in non-specific mucociliary clearance due to abnormal dehydrated mucus (Boucher 2004), but this does not explain why infection with the narrow range of unusual pathogens predominates. Optimal therapy of CF lung infections is highly controversial, some clinicians opting for early aggressive therapy at the first sign of colonization with *P. aeruginosa* or *B. cepacia* complex; other clinicians opt for therapy at the first sign of active disease.

Pneumonia in primary immunodeficiency

Chronic granulomatous disease (CGD)

CGD is a rare primary immunodeficiency, effecting about 1 in 250 000, in which phagocytic cells are unable to generate toxic ROS from oxygen (Winkelstein et al 2000). ROS are critically important elements of oxidative killing; their mechanism of action may be directly bactericidal or may be instrumental in release of toxic proteolytic enzymes (Reeves et al 2002). The disease is due to malfunction of the NADPH oxidase system as a result of a mutation in one of the four genes encoding its essential components. CGD is most commonly caused by mutations in the gene for cell surface gp91; this form is acquired by X-linked recessive inheritance and results in the most severe clinical phenotype. Milder phenotypes are acquired by autosomal recessive inheritance.

Patients with CGD are particularly susceptible to pneumonia, although soft tissue infections, osteomyelitis and liver abscesses are also frequently seen. Infections are restricted to a relatively narrow range of bacteria and fungi that are catalase-positive and/or resistant to nonoxidative phagocytic killing (the only functional bactericidal mechanism in CGD cells). The most common causes of fatal infection are *Aspergillus* species and *Burkholderia cepacia* complex; the latter are highly resistant to nonoxidative killing due to an unusual lipopolysaccharide structure.

For reasons remaining poorly explained, patients with CGD may experience serious, often life-threatening inflammatory disease of the lung, gastrointestinal tract and urinary tract (Schappi et al 2003, Kobayashi et al 2004). These inflammatory processes appear not to be caused by a cryptic infection and are highly responsive to therapy with adrenocortical steroids.

Hypogammaglobulinaemia

Congenital absence of immunoglobulins was the first primary immunodeficiency disorder described. Patients with one of the many variants of hypogammaglobulinaemia are highly susceptible to sinopulmonary infections but rarely experience invasive infectious diseases (Ballow 2002). Infections first appear after passively acquired maternal IgG is cleared from the baby's circulation—in the first year of life. Infections are caused by bacteria of relatively low virulence, such as non-typeable *Haemophilus influenzae*. Therapy with regular infusion of intravenous immunoglobulin is a highly effective means of preventing these recurrent infections.

Complement deficiency

Patients with absent or very low levels of serum complement, particularly early components of the classical pathway, are at enhanced risk of acquiring bacterial pneumonia, particularly that associated with invasive pneumococcal disease (Ekdahl et al 1995).

Job's/hyper IgE syndrome

Pneumonia, grossly elevated IgE and abscesses are the common features of Job's/hyper IgE syndrome. The condition is autosomal dominant and is often associated with skeletal abnormalities and delayed dentition. Pneumonia is most commonly caused by *H. influenzae* or *S. aureus* and can commonly lead to pneumatocele formation (Grimbacher et al 1999).

Pneumonia in acquired immunodeficiency

A wide range of immunocompromizing diseases and therapies exist which are beyond the scope of this brief chapter to review. However, such conditions as haematopoietic malignancies, therapies for cancer and immunosuppressive therapies each enhances the likelihood of acquiring pneumonia with a common or unusual pathogen. Therapy should be determined by the gravity of the disease, the pattern of the pneumonic infiltration and the potential pathogen(s) identified.

Acquired immunodeficiency syndrome (AIDS)

Individuals infected with HIV are at greatly enhanced susceptibility to pneumonia (Boyton 2005), particularly if their CD4 counts are depressed. In San Francisco County, California, patients with HIV infection were 46-fold more likely to contract pneumococcal disease than those without HIV infection (Nuorti et al 2000). Tuberculosis frequently accompanies HIV infection and is one of the AIDS-defining infections. Co-infection with HIV and *M. tuberculosis* carries a very grave prognosis if the latter is not aggressively treated (Boyton 2005). Opportunistic pulmonary infections in patients with AIDS clearly illustrate the important role played by T lymphocytes in the control of respiratory tract bacterial infection.

Strategies for preventing pneumonia

Immunization

S. pneumoniae continues to be a common cause of community-acquired pneumonia in children and adults. Immunization with either the childhood vaccine (seven-valent conjugated) or the adult formulation (23-valent unconjugated) decreases the incidence of pneumococcal infection (Dear et al 2003, Lucero et al 2004). In children who have been vaccinated against pneumococcal infection, the frequency of pneumococcal pneumonia is dramatically decreased and that of pneumonia in general is diminished by about 30%. The conjugated vaccine is now part of the standard infant immunization regimen in many parts of the developed world, and the unconjugated vaccine is recommended for older adults, for all people with serious underlying diseases that enhance their risk from pneumonia and in health care workers. Protection is not perfect and is limited to those serotypes contained in the multi-valent immunogen. Immunization against *H. influenzae* type b has virtually eliminated that organism as a cause of serious invasive disease in North America and other parts of the world (Watt et al 2003) but does not protect against infection with non-typeable or non-type b organisms. Immunization is obviously limited to those who can mount a protective response; agammaglobulinemic individuals are dependent upon passive immunotherapy with intravenous immunoglobulin. Immunization against *P. aeruginosa* has failed to prevent infection in patients with CF, although novel strategies continue to be investigated.

Immunomodulation

Patients with CGD who receive thrice weekly injections of interferon (IFN)γ appear to be protected against serious infection including pneumonia (Gallin 1991). The mechanism for this protection is not yet understood. Intravenous immunoglobulin appears to modulate immunity by mechanisms in addition to simple antibody repletion. It is effective in a number of conditions, such as idiopathic thrombocytopenic purpura and Kawasaki disease, but its mechanism of action, like that of IFNγ, is not understood. Cytokine therapy and cytokine blockade have been considered or instituted for therapy of serious bacterial infection, but none has yet been approved for general clinical care (Standiford & Deng 2004). In fact, therapy for certain inflammatory diseases with antibodies against tumour necrosis factor (TNF) have enhanced the risk of re-activation tuberculosis (Dinarello 2003).

Acknowledgments

Supported by grants from the Canadian Institutes for Health Research and the Canadian Cystic Fibrosis Foundation. I thank Drs Stuart Turvey and Anne Junker for their critical review of the manuscript.

References

Ballow M 2002 Primary immunodeficiency disorders: antibody deficiency. J Allergy Clin Immunol 109:581–591

Boucher RC 2004 Relationship of airway epithelial ion transport to chronic bronchitis. Proc Am Thorac Soc 1:66–70

Bowdish DM, Davidson DJ, Scott MG, Hancock RE 2005 Immunomodulatory activities of small host defense peptides. Antimicrob Agents Chemother 49:1727–1732

Boyton RJ 2005 Infectious lung complications in patients with HIV/AIDS. Curr Opin Pulm Med 11:203–207

Cook DN, Pisetsky DS 2004 Toll-like receptors in the pathogenesis of human disease. Nat Immunol 5:975–979

Cowan MJ, Gladwin MT, Shelhamer JH 2001 Disorders of ciliary motility. Am J Med Sci 321:3–10

Currie AJ, Speert DP, Davidson DJ 2003 *Pseudomonas aeruginosa*: role in the pathogenesis of the CF lung lesion. Semin Respir Crit Care Med 24:671–680

Dear K, Holden J, Andrews R, Tatham D 2003 Vaccines for preventing pneumococcal infection in adults. Cochrane Database Syst Rev CD000422

Dinarello CA 2003 Anti-cytokine therapeutics and infections. Vaccine 21 Suppl 2:S24–34

Ekdahl K, Truedsson L, Sjoholm AG et al 1995 Complement analysis in adult patients with a history of bacteremic pneumococcal infections or recurrent pneumonia. Scand J Infect Dis 27:111–117

Gallin JI 1991 Interferon-gamma in the treatment of the chronic granulomatous diseases of childhood. Clin Immunol Immunopathol 61(2 Pt 2):S100–105

Grimbacher B, Holland SM, Gallin JI et al 1999 Hyper-IgE syndrome with recurrent infections—an autosomal dominant multisystem disorder. N Engl J Med 340:692–702

Happel KI, Bagby GJ, Nelson S 2004 Host defense and bacterial pneumonia. Semin Respir Crit Care Med 25:43–52

Hawn TR, Verbon A, Lettinga KD 2003 et al A common dominant TLR5 stop codon polymorphism abolishes flagellin signaling and is associated with susceptibility to legionnaires' disease. J Exp Med 198:1563–1572

Hawn TR, Verbon A, Janer M 2005 Toll-like receptor 4 polymorphisms are associated with resistance to Legionnaires' disease. Proc Natl Acad Sci USA 102:2487–2489

Kobayashi SD, Voyich JM, Braughton KR et al 2004 Gene expression profiling provides insight into the pathophysiology of chronic granulomatous disease. J Immunol 172:636–643
Leijh PC, van Zwet TL, van Furth R 1984 Extracellular stimulation by serum proteins required for maximal intracellular killing of microorganisms by mouse peritoneal macrophages. Infect Immun 46:754–758
Lucero MG, Dulalia VE, Parreno RN et al 2004 Pneumococcal conjugate vaccines for preventing vaccine-type invasive pneumococcal disease and pneumonia with consolidation on x-ray in children under two years of age. Cochrane Database Syst Rev CD004977
Lynch JP 3rd 2001 Hospital-acquired pneumonia: risk factors, microbiology and treatment. Chest 119(2 Suppl):373S–384
Mason CM, Ali J 2004 Immunity against mycobacteria. Semin Respir Crit Care Med 25:53–61
Masten BJ 2004 Initiation of lung immunity: the afferent limb and the role of dendritic cells. Semin Respir Crit Care Med 25:11–20
McWilliam AS, Nelson D, Thomas JA, Holt PG 1994 Rapid dendritic cell recruitment is a hallmark of the acute inflammatory response at mucosal surfaces. J Exp Med 179:1331–1336
Nuorti JP, Butler JC, Gelling L, Kool JL, Reingold AL, Vugia DJ 2000 Epidemiologic relation between HIV and invasive pneumococcal disease in San Francisco County, California. Ann Intern Med 132:182–190
Reeves EP, Lu H, Jacobs HL et al 2002 Killing activity of neutrophils is mediated through activation of proteases by K+ flux. Nature 416:291–297
Schappi MG, Klein NJ, Lindley KJ et al 2003 The nature of colitis in chronic granulomatous disease. J Pediatr Gastroenterol Nutr 36:623–631
Segreti J, House HR, Siegel RE 2005 Principles of antibiotic treatment of community-acquired pneumonia in the outpatient setting. Am J Med 118(Suppl 7A):21S–28
Shaw MJ 2005 Ventilator-associated pneumonia. Curr Opin Pulm Med 11:236–241
Speert DP 2002 Advances in *burkholderia cepacia* complex. Paediatr Respir Rev 3:230–235
Speert DP, Bond M, Woodman RC, Curnutte JT 1994 Infection with Pseudomonas cepacia in chronic granulomatous disease: role of nonoxidative killing by neutrophils in host defense. J Infect Dis 170:1524–1531
Standiford TJ, Deng JC 2004 Immunomodulation for the prevention and treatment of lung infections. Semin Respir Crit Care Med 25:95–108
Twigg HL 3rd 2004 Macrophages in innate and acquired immunity. Semin Respir Crit Care Med 25:21–31
Wang Q, Doerschuk CM, Mizgerd JP 2004 Neutrophils in innate immunity. Semin Respir Crit Care Med 25:33–41
Watt JP, Levine OS, Santosham M et al 2003 Global reduction of Hib disease: what are the next steps? Proceedings of the meeting Scottsdale, Arizona, September 22–25, 2002. J Pediatr 143(6 Suppl): S163–187
Winkelstein JA, Marino MC, Johnston RB Jr et al 2000 Chronic granulomatous disease. Report on a national registry of 368 patients. Medicine (Baltimore) 79:155–169

DISCUSSION

Hussell: We work on the CyBB knockout mouse that is missing gp91phox. It is identical to the scenario you present in that we get excessive production of innate inflammatory cytokines when the ability to produce reactive oxygen and nitrogen species is missing. It seems to be due to an inherent defect in turning off the macrophage. It is almost as if these low level innate signals are needed to instruct a macrophage to rest which have been taken away, and homeostasis is lost. These patients have heightened macrophage responses. Is that where your inflammatory cytokines are coming from?

Speert: The macrophage is not the only cell that expresses NADPH oxidase. Neutrophils and monocytes are probably important sources of proinflammatory cytokines.

Hussell: We find they clear infections better, because the macrophages are in a heightened activation state. Viral infections, at least, are cleared better.

Speert: This is in a gp91phox knockout mouse. That is another example of the disparity one sees between human and murine disease.

Sheehan: Do your primary ciliary dyskinesia (PCD) patients cough a lot?

Speert: Not like CF patients. They don't have a problem with non-specific mucociliary clearance mechanisms. They will cough if they have pneumonia, but this is usually relatively short-lived.

Sheehan: So there are other forms of PCD, which do have cough?

Speert: In Kartagener's syndrome or other ciliary dyskinesia conditions, patients do cough. This would be expected. It is the same thing with children who have had pertussis. It is called 100 day cough because they lose their cilia for 100 days. In the meantime the only mechanism for moving trapped organisms back to the mouth is by coughing. Ciliary clearance is very important.

Brown: You said that the chronic granulomatous disease (CGD) heterozygote patients have some kind of obstructive disorder?

Speert: The patients with CGD do have problems with hyperinflammation irrespective of the genotype. The carrier mothers usually have about 50% normal neutrophils and 50% abnormal cells, and they have problems with a lupus-like illness. All the patients with CGD have problems with inflammation, some more than others. There is one sibship of two brothers whom I have treated, one who has never had infection and is now being prepared for lung transplantation because of inflammatory lung disease. The other has had multiple infections and no problems with inflammation. Both, however, share the same mutation. There are clearly other genes and regulatory mechanisms that determine whether this disease will be primarily infectious versus inflammatory.

Brown: You said that inflammation is related to gp91. How?

Speert: The way we approach things in the lab is that once we have a clinical phenotype, we look for a cellular phenotype that can then be studied. The cellular phenotype in our patients with CGD is excessive production of all NF-κB-dependent inflammatory cytokines, so we have now been studying this at the molecular level, to see if we can determine whether there is something in the NF-κB signalling cascade that is abnormal. We have looked at whether NF-κB is being translocated normally and whether I-κB is degraded normally. I-κB is degraded normally and phosphorylated normally. p50 translocates entirely normally, but we never see p65 in the nucleus. This suggests that there is a bizarre transcriptome. We have demonstrated this by Western analysis, EMSA and supershifting. No p65 gets into the nucleus. There is some *trans*-activating component of the NF-κB complex that is driving hyperinflammation, but we have not figured it out yet.

Brown: What is happening to gp91?

Speert: It is related to a defect in production of ROS. The ROS are involved in signalling, but the mechanism hasn't been determined. But they have far more to do than simply kill. From the array of clinical manifestations these patients display, it is clear that a lot more is going on than simple failure to kill bacteria. ROS may have a role in regulation of inflammation. The first thing we need to work out is the partner for p50 in the nucleus.

Ryffel: Does the p90 knockout also have an inflammatory syndrome?

Speert: Gp91phox deficiency and p47phox deficiency both result in a hyperinflammatory state. It doesn't matter what the mutation is: if there is diminished production of ROS, the result is a hyperinflammatory state.

Ryffel: In the CGD-deficient mouse, is IFNγ also active?

Romani: Yes.

Speert: To follow up on that, Vancouver has a large Asian population. Out of the 12 patients with CGD, two are Asian. They are the only two patients in our clinic who have developed lupus. Asian patients are much more likely to develop lupus than Caucasians. There may be an overwhelming amount of cellular or other debris that builds up in the system that has to be cleared. If you exceed that threshold, an autoimmune condition might occur. There is something about being Asian that lowers the threshold for autoimmune diseases, and CGD lowers the threshold even further.

Mantovani: With IFNγ treatment, did you look at responsiveness of cells from patients treated this way?

Speert: That is a great question. We tried very hard to find even subtle differences between patients on IFNγ or off it. We can't find any clear differences. Of the two siblings I mentioned with the different phenotypes, one is on IFNγ and the other isn't, and their cells look exactly the same *in vitro*. Every patient we have looked at translocates p50 but not p65. In every one, there is a hyperinflammatory condition.

Mantovani: A few years ago we found that reactive oxygen intermediates cause rapid release of the IL1 decoy receptor from monocytes. Possibly, the monocyte could be denuded and could be more sensitive.

Speert: We have stimulated with IL1 and they are hyperresponsive. Do you think if we used serum from these patients, if we had the decoy present it would block this phenomenon?

Mantovani: Yes.

Segal: Have you checked whether these agonists that you use for the Toll receptors switch on the oxidase or not, in normal cells? If they do, if you are comparing normal with abnormal cells all sorts of things are happening in terms of metabolic changes and ion fluxes, which could be influencing the results and which are unrelated to oxygen radical production. The effects that you see might be related to the metabolic changes that are taking place in these cells.

Speert: I don't know the answer to that. But assume what you say is the case, can you then explain the inability to find a partner to p50?

Segal: No, but if it is a monogenic disease, it is unlikely to be due to the fact that another protein is lacking.

Speert: p65 is there, and we can find it in normal levels in the cytoplasm.

Segal: The conditions in the cytoplasm might be different in the two types of cells.

Lambrecht: Have you looked at haem oxygenase levels in those patients? Haem oxygenase 1 (HO1) regulates the enzyme 2,3-indoleamine deoxygenase (IDO) levels. IDO enzyme activity is broadly anti-inflammatory. IFNγ is one of the inducers of the IDO enzyme. There is a lot of oxidative damage and if HO1 goes down, then IDO will go down as well.

Speert: As I understand it, IFNγ doesn't enhance reactive oxygen radical production in CGD cells.

E Sim: Do the non-steroidal anti-inflammatory drugs (NSAIDs) work either in the mouse or the human? The reason I am asking is that there might be a biochemical effect that is related to lipid metabolism. This may in turn be acting as a mediator.

Speert: NSAIDs may partially control the lupus-like illness in these patients, but the classical rheumatologists would tell you that the disease these patients get is sufficiently different from classical lupus that they have a hard time even calling it systemic lupus.

Hussell: I am not familiar with anti-inflammatory drugs in the mouse. I know you can turn the macrophage off by giving it positive signal, which is what it is lacking when it hasn't got NO and superoxide species. It needs that negative feedback loop to turn itself off. If you give it IFNγ it will turn the macrophage off.

Steyn: I would like to go back to the pathogenesis of the bacterial infection. My understanding was that viruses can cause an unmasking of attachment sites on cells. Attachment is a prerequisite for any infectious process; would we have to invoke immune suppression if that is the mechanism involved?

Speert: This would hold for respiratory coinfections, such as influenza preceding pneumococcal infections. It wouldn't explain the observations in chicken pox, where the onset of the prototypic bacterial infection (with group A streptococci) has its onset on day five of the illness, at a time when one would think the individual would be totally protected as most of the skin lesions have crusted and the patient is nearly returned to full health. There are animal models that show that major infections result in transient neutrophil dysfunction at days 3 and 4. This effect could be important, but I think it is more complex than that.

Feldman: I'd like to focus on the chicken pox pneumonia. We have had an experience in adults with chicken pox pneumonia. We have a cohort of HIV– and more recently a cohort of HIV+ patients with chickenpox pneumonia. Particularly in adults who smoke, we have found a group who go on to develop what is called chicken pox pneumonia. In our experience, the pneumonia doesn't appear to be simply a secondary bacterial infection. It appears to be almost like an immunological response. We have treated these patients with the additional use of corticosteroids. We would use an antiviral agent and an antibiotic routinely, but the most important component of therapy has been a corticosteroid to reverse that diffuse pulmonary infiltrate that develops. This seems to be more significant than secondary bacterial infection. We have managed to reverse the pulmonary infiltrates even in severe cases within 24 h and avoided the need for ventilation in patients who were heading that way simply by the additional use of steroids.

Speert: Chicken pox is a nasty disease to get as an adult. The timing is important in trying to determine the aetiology. Chicken pox is notorious for causing non-infectious host inflammatory disease, particularly in the brain. I wouldn't be surprised if it could do the same in the lung. What is called chicken pox pneumonia can be a terrible disease. What is the usual timing for this?

Feldman: It is more or less the kind of timing that you are talking about. Adults come in with the florid skin rash, and two or three days later develop these progressive pulmonary infiltrates, heading towards the need for ventilation. In our experience with the first cohort of patients that we studied, there is quite a mortality in adults who develop respiratory failure and end up on a ventilator in the intensive care unit, but our mortality in patients given corticosteroids was zero. In Cape Town a study has been published pushing for the use of antiviral agents. In our experience, additional steroids are more important.

I noticed that you work in a paediatric unit. The comment you made is that most pneumonias occurring in children are viral. I think in adults it is probably different: it is much more bacterial, particularly when we talk about HIV.

Speert: The aetiology of pneumonia depends on age of the patient and the underlying immunological status.

Gordon: What is the normal flora of the upper respiratory tract?

Speert: The normal commensals are the same organisms that cause infections in patients who have mucociliary dyskinesia. They are the same organisms that cause middle-ear infections in children. If there is some defect that blocks the eustacian tube or allows aspiration of upper respiratory secretions, it is the resident upper respiratory flora that causes problems. Many people carry potential pathogens in the upper respiratory tract. 25% of children carry group A streptococcus in their throat. A substantial number carry *Staphylococcus aureus* in their nasopharynx.

Pathogenesis of avian flu H5N1 and SARS

Malik Peiris

Department of Microbiology, The University of Hong Kong, University Pathology Building, Queen Mary Hospital Compound, Pokfulam, Hong Kong

Abstract. Avian influenza A (H5N1) and severe acute respiratory syndrome (SARS) coronavirus are infections that cause a severe viral pneumonia leading to acute respiratory dysfunction syndrome and carry a high case-fatality rate. We have investigated innate immune responses to both viruses using primary human macrophages and respiratory epithelial cells as *in vitro* models. In contrast to human influenza A H1N1 viruses, the H5N1 viruses hyper-induce cytokines (tumour necrosis factor [TNF]α, interferon β) and chemokines (IP10, MIP1α, MCP) in *in vitro* cultures of primary human macrophages. A similar differential effect is observed in primary human bronchial epithelial cells and in type 2 pneumocytes although TNFα is not induced in respiratory epithelial cells. The cell signalling pathways responsible for this differential effect remain to be explored. Preliminary data suggest that such differential signalling involves p38 MAP kinase rather than NF-κB. SARS coronavirus infection of primary human macrophages is associated with a strong induction of chemokines without an associated type 1 interferon response. These observations may be relevant in disease pathogenesis.

2006 Innate immunity to pulmonary infection. Wiley, Chichester (Novartis Foundation Symposium 279) p 56–65

Avian flu (H5N1) and severe acute respiratory syndrome (SARS) are two infectious diseases that have recently emerged to threaten human health. While these two diseases are caused by very different viruses, highly pathogenic avian influenza (HPAI) virus H5N1 and a hitherto unknown coronavirus respectively, these diseases share a number of important similarities. Both arose from an animal reservoir and have presumably crossed zoonotically to humans repeatedly, over many years. SARS coronavirus adapted to efficient human transmission in 2002 and spread rapidly across the world (Osterhaus & Peiris 2005). The avian influenza (H5N1) virus not yet achieved efficient transmissibility in humans and remains zoonotic. Whether it will ever do so remains unclear. However, the experience of SARS is a stark reminder that such a possibility cannot be excluded, although the probability or timing of such an event remains uncertain. Both diseases give rise to a fulminant viral pneumonia, rapidly progressing to acute respiratory distress syndrome (ARDS) and are associated with significant case-fatality. At autopsy, massive macrophage infiltration and diffuse alveolar damage are seen in the lung in both diseases. However, the mechanisms that underlie disease pathogenesis remains unclear.

Avian influenza A (H5N1)

The clinical features and pathogenesis of human H5N1 disease are fundamentally different from those of conventional human influenza (Beigel et al 2005). It affects healthy children or young adults, leads to a rapidly progressive viral pneumonia, the disease severity is independent of secondary bacterial infection, leads to ARDS and is associated with multiple organ dysfunction.

The viral determinants of pathogenesis of HPAI disease in chicken are well understood. The viral haemagglutinin (HA) has to undergo post-translational cleavage into HA1 and HA2, exposing the fusogenic domain of the HA2 that mediates fusion between the membranes of the viral envelope and lysosomal membrane. This cleavage is mediated by host proteases. Low pathogenic influenza viruses (LPAI) have a single arginine at the HA1–HA2 cleavage site and are cleaved by trypsin-like proteases which have a limited tissue distribution, being found in the respiratory and gastro-intestinal tract epithelia. However, influenza subtypes H5 and H7 may acquire multiple basic amino acids at the HA1–HA2 connecting peptide which allows a range of ubiquitous furin-like proteases to cleave the HA molecule (Horimoto & Kawaoka 2005). The wide tissue distribution of these proteases implies that these HPAI viruses can replicate in multiple organs including the brain. HPAI in chicken is a disseminated and fulminant disease, often associated with dissemination to the brain and leading to rapid and sudden death in poultry. How relevant the H5 multi-basic cleavage site is in the pathogenesis of human H5N1 disease is unclear. It clearly contributes in part, to dissemination and pathogenesis of H5N1 virus in mice (Hatta et al 2001). In humans, the H5N1 virus has been found to disseminate beyond the respiratory tract to affect the brain in occasional patients (de Jong et al 2005). However, in the majority of patients, severe disease and death is primarily attributable to the respiratory illness rather than to viral dissemination. The unusual severity of the respiratory disease and the pathogenesis of ARDS in humans remains to be explained. Patients with H5N1 disease have evidence of reactive haemophagocytosis affecting multiple organs including the bone marrow, lymph nodes, spleen and even the meninges (To et al 2001, Peiris et al 2004). This may be a reflection of hyper-cytokinaemia. When compared with human influenza A H3N2 infection, patients with H5N1 disease have higher levels of chemokines (e.g. IP10) in their serum (Peiris et al 2004). Whether this is simply a consequence of more extensive lung pathology or whether these high cytokine levels contribute to the unusual pathogenesis was not clear. *In vitro* studies on primary human respiratory epithelium infected with H5N1 and human influenza viruses (H1N1) indicate that H5N1 viruses are associated with differentially enhanced chemokine and cytokine (IP10, RANTES, IL6, interferon β) responses (Chan et al 2005). Similarly, when compared with human influenza viruses, primary human macrophages infected with H5N1 viruses differentially hyper-induce pro-inflammatory cytokine (e.g. TNFα, interferon β) and chemokine responses (IP10, RANTES, MIP1α, MIP1β) (Cheung et al 2002). We have therefore hypothesized that H5N1 infection of the respiratory epithelium leads to enhanced chemokine responses that lead to the attraction of macrophages into the lung. The spill over of H5N1 infection into macrophages may then lead to a massive pro-inflammatory cytokine cascade and lead to ARDS. The mechanisms associated with such cytokine hyper-induction by the H5N1 viruses are unclear. However, p38 MAP kinase signalling seems to be differentially activated by H5N1 and this may be one signalling pathway that is specifically hyper-activated by the H5N1 virus (Lee et al 2005). Defining such pathways may allow targeted interventions and may potentially lead to improved therapy for patients with H5N1 disease.

The emergence of a pandemic requires the H5N1 virus to acquire the ability for sustained human-to-human transmission. Given the segmented RNA genome of influenza viruses, this may occur either through genetic reassortment with a human influenza virus or through adaptation through mutation. It appears that the property for cytokine hyperinduction is not determined primarily by the haemagglutinin. Therefore one may speculate that a pandemic virus that emerges via reassortment with a human influenza viruses may be associated with markedly less severe disease than seen with current H5N1 viruses. However, a H5N1 virus that acquires human-to-human transmissibility via mutation while retaining its full 'avian-like' gene complement may retain much of its disease severity—a distinctly unpleasant prospect.

SARS

The viral load of SARS coronavirus in the upper respiratory tract is low early in the disease and progressively increases to peak at around day 10 of illness (Peiris et al 2003, Chan et al 2004). This is associated with the observation that transmission of SARS is less likely in the first 5 days of illness, an epidemiological feature of the disease that permitted public health interventions (case detection and isolation) to have such dramatic success in interrupting disease transmission. The viral load appears to start to decline after about day 10 of illness, around the time of the appearance of the adaptive immune response (e.g. neutralizing antibody). We considered whether this viral load profile was an indication of defects in innate immune function (Lau & Peiris 2005). Genetic polymorphism studies have demonstrated an association between disease susceptibility and genotypes associated with low levels of mannose binding lectin in the serum (Ip et al 2005).

Patients with SARS coronavirus infection have elevated levels of serum pro-inflammatory cytokines (IL1, IL6, IL12) and chemokines (e.g. IP10, MCP1, IL8) in their serum (Wong et al 2004). Infection of epithelial cells, macrophages and dendritic cells *in vitro* with SARS coronavirus is associated with induction of chemokines such as IP10, MIP1α, MCP1. However, there is a notable lack of type 1 interferon or TNFα responses (Cheung et al 2005, Spiegel et al 2005, Law et al 2005). Interestingly, interferon response genes appear to be activated in spite of the deficient type 1 interferon response. However, the lack of type 1 interferon from macrophages and dendritic cells which otherwise helps protect adjacent cells from infection may afford the SARS coronavirus an opportunity to evade innate immune responses. These *in vitro* data are corroborated by a study of gene expression profiles of peripheral blood mononuclear cells from SARS infected patients compared with patients with influenza. The patients with SARS had little or no induction of type 1 interferon responses in their peripheral blood leukocytes in contrast to patients infected with influenza A (Reghunathan et al 2005).

In conclusion, both SARS and avian flu H5N1 lead to diffuse alveolar damage in the lung contributing to the severe disease outcome. The mechanisms involved are still unclear. However, comparison of the innate immune responses in these two infections may provide clues relevant to pathogenesis.

References

Beigel JH, Farrar J, Han AM et al 2005 Avian influenza A (H5N1) infection in humans. N Engl J Med 353:1374–1385

Chan KH, Poon LL, Cheng VC et al 2004 Detection of SARS coronavirus in patients with suspected SARS. Emerg Infect Dis 10:294–249

Chan MC, Cheung CY, Chui WH et al 2005 Proinflammatory cytokine responses induced by influenza A (H5N1) viruses in primary human alveolar and bronchial epithelial cells. Respir Res 6:135

Cheung CY, Poon LL, Lau AS et al 2002 Induction of proinflammatory cytokines in human macrophages by influenza A (H5N1) viruses: a mechanism for the unusual severity of human disease? Lancet 360:1831–1837

Cheung CY, Poon LL, Ng IH et al 2005 Cytokine responses in severe acute respiratory syndrome coronavirus-infected macrophages in vitro: possible relevance to pathogenesis. J Virol 79:7819–7826

de Jong MD, Bach VC, Phan TQ et al 2005 Fatal avian influenza A (H5N1) in a child presenting with diarrhea followed by coma. N Engl J Med 352:686–691

Hatta M, Gao P, Halfmann P, Kawaoka Y 2001 Molecular basis for high virulence of Hong Kong H5N1 influenza A viruses. Science 293:1840–1842

Horimoto T, Kawaoka Y 2005 Influenza: lessons from past pandemics, warnings from current incidents. Nat Rev Microbiol 3:591–600

Ip WK, Chan KH, Law HK et al 2005 Mannose-binding lectin in severe acute respiratory syndrome coronavirus infection. J Infect Dis 191:1697–1704

Lau YL, Peiris JS 2005 Pathogenesis of severe acute respiratory syndrome. Curr Opin Immunol 17:404–410
Law HK, Cheung CY, Ng HY et al 2005 Chemokine up-regulation in SARS-coronavirus-infected, monocyte-derived human dendritic cells. Blood 106:2366–2374
Lee DC, Cheung CY, Law AH, Mok CK, Peiris M, Lau AS 2005 p38 mitogen-activated protein kinase-dependent hyperinduction of tumor necrosis factor alpha expression in response to avian influenza virus H5N1. J Virol 79:10147–10154
Osterhaus ADME, Peiris M 2005 Lessons learnt. In: Peiris M, Anderson LJ, Osterhaus ADME, Stohr K, Yuen KY (eds) Severe acute respiratory syndrome. Blackwells, Oxford, p 249–253
Peiris JS, Chu CM, Cheng VC et al 2003 Clinical progression and viral load in a community outbreak of coronavirus-associated SARS pneumonia: a prospective study. Lancet 361:1767–1772
Peiris JS, Yu WC, Leung CW et al 2004 Re-emergence of fatal human influenza A subtype H5N1 disease. Lancet 363:617–619
Reghunathan R, Jayapal M, Hsu LY et al 2005 Expression profile of immune response genes in patients with severe acute respiratory syndrome. BMC Immunol 6:2
Spiegel M, Pichlmair A, Martinez-Sobrido L et al 2005 Inhibition of Beta interferon induction by severe acute respiratory syndrome coronavirus suggests a two-step model for activation of interferon regulatory factor 3. J Virol 79:2079–2086
To KF, Chan PKS, Chan KF et al 2001 Pathology of fatal human infection associated with avian influenza A H5N1 virus. J Med Virol 63:242–246
Wong CK, Lam CW, Wu AK et al 2004 Plasma inflammatory cytokines and chemokines in severe acute respiratory syndrome. Clin Exp Immunol 136:95–103

DISCUSSION

Schoub: Can you comment on the potential effect of tropism with α2,3 and α2,6? We know that in birds the α2,3 receptor is utilized by avian influenza virus. We also know that in the human H5N1 infections there is quite a substantial gastro-intestinal component, with most of the patients presenting with watery diarrhoea. The virus still retains the α2,3 receptor in human infections. The 1918 virus seemed to utilize the human α2,6 receptor. Could the tropism of the virus have some relationship to the increased virulence, in that it may play a role in the up-regulation of cytokines? In addition to the viral component you may get a host component as well.

Peiris: You can think about the receptor bias of the α2,3/2,6 effect at two levels: first of all, initial susceptibility to infection, and second, the severity of disease. If you think about the number of individuals across Asia who have been exposed to the HN51 virus, we are talking about tens of thousands of individuals. Only a tiny proportion of these people have become sick. There is a huge discrepancy. The exposure to the pathogen is necessary but not sufficient. There is a huge barrier at the moment in terms of exposure and disease. Whether this is due to receptor genetic polymorphisms in individuals it is unclear.

In terms of severity of disease, receptors may also play a role, for example, in terms of the receptor distribution in certain individuals and tissues. Is it more α2,3 than α2,6 linked sialic acids at different levels of the respiratory tract and is this the same in all individuals? However, at the individual cell level we still come back to the *in vitro* observations that the H5N1 virus, in the same cells from the same individual, is behaving fundamentally differently to the human flu (H1N1 or H3N2) virus. In addition to what role the receptor may be playing, the H5N1 virus behaves differently, at least in its outcome on the host cell interaction.

Schoub: The reason I ask is that if it does change, and mutates to α2,6 receptor binding, could this relate to a dilution of pathogenesis?

Peiris: This is an assumption made by some people. They think the adaptation of the virus for it to be able to transmit from human–human would lead to a diminution of its virulence. This is not really based on fact. If the virus reassorts then this could change virulence significantly. But if the virus adapts to transmit efficiently from human–human by mutation of its haemagluttinin, there is no reason why this change by itself would affect pathogenesis.

Schoub: Unless it is receptor-related pathogenesis.

Peiris: I think that is stretching it.

Steyn: Workers at the US Armed Forces Institute of Pathology have just recreated the 1918 influenza virus (Tumpey et al 2005). On the basis of studies *in vitro* with this virus they have shown two things: one is that the DNA polymerase is more effective than normal flu viruses; and, second, this virus is infectious in the absence of trypsin. This implies that the haemagluttinin doesn't have to be cleaved proteolytically for the virus to be infectious.

I have another point, with reference to TB. There is a TB strain, CD1551, which is also very effective at inducing a TNFα response. The irony is that although this mycobacterium can be transmitted, it is a less virulent bacterium because the immune system effectively forms granulomas which contain the organism.

Steinman: I thought influenza was supposed to be a cytopathic virus. I don't recall your kinetics, but the monocytes didn't die right away. For us, human monocytes die in 12 h. Are we doing something wrong?

Peiris: They do die, eventually.

Steinman: It seems very slow.

Peiris: The virus replicates, and just like in epithelium, it is cytopathic and kills the cells. But in the meantime, it is putting out all these cytokines and chemokines.

Steinman: Is the time required to see the cytopathic effect the same as H1N1?

Peiris: No, with H5N1 it is slower. Ultimately the cells do die, but there seems to be a delay in apoptosis in human cells. This is true also for some other avian viruses in human cells.

Finn: This may have to do with p38. You showed that this was enhanced, which prolongs the life of a cell.

Steinman: You didn't show interferon production in H5N1 infections: is there anything unusual there?

Peiris: In addition to TNFα, H5N1 viruses also strongly up-regulate interferon β. One could argue that this should be protecting the host and stopping the viral replication. Rob Webster has shown that H5N1 viruses are more resistant to the antiviral effects of interferon (Seo et al 2002) than normal human flu viruses. Those data come from studies on pig epithelial cells, however. We have looked in the human epithelial cell system and don't see such a big difference. One would expect that the interferon should be protective to some extent.

Lambrecht: You mentioned briefly that particularly in the H5N1 model, type 2 pneumocytes are infected by virus. These cells are highly involved in lung fibrosis pathogenesis. Do patients who survive infection progress into lung fibrosis, which has protected them?

Peiris: This has not really been a feature of patients who survive H5N1. It has certainly been a feature of patients who survive from severe acute respiratory syndrome (SARS). They clearly have a fibro proliferative phase, but this hasn't been clearly documented in H5N1.

Lambrecht: So SARS also goes to the pneumocytes.

Peiris: There are a number of different groups who have published that the pneumocyte is infected (Nicholls et al 2006). There are also *in vitro* data suggesting that if SARS coronavirus infects differentiated human ciliated airway epithelium (Sims et al 2005), the undifferentiated epithelium is not infected by SARS coronavirus. It seems likely that a major target *in vivo* would be pneumocytes.

Gordon: Is the receptor the ACE2?

Peiris: In SARS we don't actually see ACE2 in the undifferentiated pneumocytes or respiratory epithelial cells, but when respiratory epithelium is differentiated *in vitro*, ACE2 is detectable (Sims et al 2005). We have tried to infect the human undifferentiated respiratory epithelium and pneumocytes *in vitro* with SARS coronavirus and couldn't demonstrate infection.

Ryffel: You said there is a chemokine storm, and macrophages are recruited. But what is the cause of death of SARS virus infection? Could we learn anything from the animal model?

Peiris: The animal models can only go so far. They do not replicate the disease seen in humans in terms of the slow progression. One has to use high challenge doses to get a reproducible animal model going and in such circumstances, the pathology peaks at around 4 or 5 days. After that, the animals are recovering or they die. In comparison to the human disease progression, the course of the disease in experimental animal models is highly abbreviated. I think one may learn something from this, but I am not sure if the information can be transposed directly to the human situation.

Didierlaurent: You didn't mention T cell responses. Do you see any T cells in your pathological sections from patients with H5N1 disease?

Peiris: In H5N1 autopsy tissue, T cell infiltrates are not that prominent.

Didierlaurent: Do patients with H5N1 have a genetic problem linked to T cell responses?

Peiris: They don't necessarily have a prior history of susceptibility to previous viral infections. These people seem to be perfectly healthy prior to H5N1 infection. Of course one cannot exclude a subtle genetic defect that defines a key response that is peculiar to this virus.

Didierlaurent: Does the fact that they have different innate responses affect the T cell response?

Peiris: That is certainly possible.

Williams: Do you have a neutrophil accumulation as well as monocyte accumulation? None of your chemokines were neutrophil attractants.

Peiris: In autopsy tissues, neutrophils are not prominent. However, one doesn't know what happens very early in disease. Certainly, by the time they die, it is predominantly monocytes and macrophages rather than neutrophils that are found in the lung. This is different to typical adult respiratory distress syndrome (ARDS) which follows septic shock where lots of neutrophils are seen in the lungs. Neither SARS nor avian flu have marked involvement of neutrophils, at least as seen in autopsy.

Lambrecht: Just having macrophages in your lungs by themselves is relatively harmless. There are some lung diseases where the lung becomes filled with macrophages (e.g. desquamative interstitial pneumonia). Yet these patients are restored with a shot of steroids. It is strange that these macrophages lead to death. This is when they are superactivated.

Peiris: It is likely that the macrophages aren't just sitting there, but are also getting infected by the virus. This triggers a further cascade of cytokines.

Romani: Does the avian flu virus infect myeloid and plasmacytoid DCs differently?

Peiris: We are beginning to look at the myeloid DCs, so we don't know yet.

Romani: What is your guess?

Peiris: Ask me in a year.

McGreal: Would these patients benefit from anti-TNFα treatment?

Peiris: That would be worth investigating.

Hussell: We are trying to generate funding to look at this, with a shorter-acting TNF inhibitor with a half-life of five or six hours.

Peiris: Steroid use has been tried in H5N1 patients, and has no obvious benefit. This is bearing in mind that the key with steroid use is knowing when to give it. If you have something more selective than steroids, that is likely to be a better bet than trying to use a blunt instrument like steroids.

Gordon: Can you comment on anti-retrovirals in the context of H5N1 and SARS?

Peiris: Initially, a whole range of currently available antivirals were screened for activity against SARS. One of the compounds that did have activity was lopinavir/ritonavir. No controlled clinical trial data are available. But, in a retrospective analysis of patients in Hong Kong where in the first phase of the outbreak, patients were treated with ribavirin while in the second phase some of the patients were treated with lopinavir/ritonavir in addition to ribavirin, the mortality rates of patients treated with lopinavir/ritonavir plus ribavirin had markedly improved outcome. However, the interpretation of studies with retrospective controls has to be treated with caution.

For H5N1 virus, the antivirals used have been amantadine and the neuraminidase inhibitor oseltamivir (Tamiflu). The viruses that are now causing disease in Thailand and Vietnam are resistant to amantadine, so oseltamivir is the only drug that could be expected to have any effect. The problem is that by the time many of the patients are diagnosed, they are already several days into the illness and have rapidly progressive pneumonia and perhaps ARDS. By this time, you wouldn't expect success with antivirals. Worryingly, there are some patients who have started treatment early with Tamiflu and still haven't responded.

Schoub: Relenza has also been shown to be effective. Tamiflu is not registered in this country (South Africa) yet. We have stockpiled a small quantity of Relenza for laboratory staff. My understanding is that Relenza might have a marginal advantage. The only demonstrated resistance to Tamiflu is a tyrosine mutation at position 274 which is not seen with Relenza.

Peiris: The problem with Relenza which is given by inhalation, is that it won't be ideal to treat a virus like this which is causing disease deep in the lung and which has the potential to spread systemically. The second concern is how to administer this drug by inhalation to patients who are seriously ill and have respiratory distress. There is also a question as to how well an orally administered drug like oseltamivr is absorbed in a severely ill patient.

Gordon: You implied that H5N1 is primarily a gastrointestinal infection. How does it get into the airway? Is it not spread via droplets? Is it spread by food and handling only?

Peiris: The route of the initial infection to humans is unclear. It is through handling dead poultry, so it could be by droplets contaminating the respiratory tract or conjunctiva or even by ingestion. There are at least a couple of cases where ingestion is the route. Two Vietnamese brothers ate a special traditional dish that was topped with fresh duck's blood. This is the presumed source of infection for these two.

Lambrecht: Cats and tigers are also sensitive to H5N1. Albert Osterhaus did an experiment that involved infecting chickens and feeding them to cats. The cats were infected by the GI route.

Peiris: Similarly, there was an outbreak in a zoo in Thailand. At least in animals, the oral route seems to be one way the virus is transmitted.

References

Nicholls JM, Butany J, Poon LL et al 2006 Time course and cellular localization of SARS-CoV nucleoprotein and RNA in lungs from fatal cases of SARS. PLoS Med 3:e27

Seo SH, Hoffmann E, Webster RG 2002 Lethal H5N1 influenza viruses escape host anti-viral cytokine responses. Nat Med 8:950–954

Sims AC, Baric RS, Yount B, Burkett SE, Collins PL, Pickles RJ 2005 Severe acute respiratory syndrome coronavirus infection of human ciliated airway epithelia: role of ciliated cells in viral spread in the conducting airways of the lungs. J Virol 79:15511–15524

Tumpey TM, Basler CF, Aguilar PV et al 2005 Characterization of the reconstructed 1918 Spanish influenza pandemic virus. Science 310:77–80

Immunity and tolerance to *Aspergillus fumigatus*

Claudia Montagnoli, Silvia Bozza, Roberta Gaziano, Teresa Zelante, Pierluigi Bonifazi, Silvia Moretti, Silvia Bellocchio, Lucia Pitzurra and Luigina Romani[1]

Department of Experimental Medicine and Biochemical Science, University of Perugia, 06126 Perugia, Italy

Abstract. The inherent resistance to diseases caused by *Aspergillus fumigatus* suggests the occurrence of regulatory mechanisms that provide the host with adequate defence without necessarily eliminating the fungus or causing unacceptable levels of host damage. Efficient responses to the fungus require different mechanisms of immunity. Dendritic cells (DCs) are uniquely able to decode the fungus-associated information and translate it into qualitatively different T helper (Th) and regulatory (T_{reg}) cell responses. A division of labour occurred between functionally distinct T_{reg} that were coordinately activated by a CD28/B.7-dependent costimulatory pathway after exposure of mice to *Aspergillus* conidia. Early in infection, inflammation was controlled by the expansion, activation and local recruitment of $CD4^+CD25^+$ T_{reg} capable of suppressing neutrophils through the combined actions of interleukin (IL10) and cytotoxic T lymphocyte antigen 4 (CTLA4) on indoleamine 2,3-dioxygenase (IDO). The levels of IFNγ produced in this early phase set the subsequent adaptive stage by conditioning the IDO-dependent tolerogenic program of DCs and the subsequent activation and expansion of tolerogenic T_{reg}, which produced IL10 and transforming growth factor (TGF)β, inhibited Th2 cells, and prevented allergy to the fungus. Thus, regulation is an essential component of the host response in infection and allergy to the fungus, and its manipulation may allow the pathogen to overcome host resistance and promote disease.

2006 Innate immunity to pulmonary infection. Wiley, Chichester (Novartis Foundation Symposium 279) p 66–79

Aspergillus fumigatus, a thermotolerant saprophyte, is associated with a wide spectrum of diseases in humans that includes saprophytic colonization of pre-existing cavities (aspergilloma), allergic asthma, allergic bronchopulmonary aspergillosis occurring as a complication of bronchial asthma or cystic fibrosis, and invasive aspergillosis in immunocompromised patients. Immunocompetent and non-atopic subjects are relatively resistant to *A. fumigatus* diseases and disease occurs in the

[1]This paper was presented at the symposium by Luigina Romani, to whom correspondence should be addressed.

setting of host damage (Latge 1999, Marr et al 2002b). Most of the inhaled conidia are eliminated by exclusion mechanisms, which include physical barriers, such as mucus and cilia, as well as a variety of mediators of the collectin family, such as lung surfactant proteins SP-A, SP-D, mannose-binding lectins (MBL) and pentraxin 3, with antimicrobial and immunomodulatory properties (Garlanda et al 2002, McCormack & Whitsett 2002, Madan et al 2005). Patients with single nucleotide polymorphisms (SNPs) in SP-A2 and MBL genes showed significant associations with *Aspergillus* infection and allergy (Madan et al 2005). Many aspects of the antimicrobial host response are orchestrated by a complex network of cytokines and their receptors (Phadke & Mehrad 2005). Tumor necrosis factor (TNF)α and interleukin (IL)6 have been shown to be required for initiation of the innate response to the fungus (Cenci et al 2001, Phadke & Mehrad 2005). Several recruitment cytokines play critical roles in mediating influx of specific leukocytes to the site of infection in experimental pulmonary aspergillosis. Among these, a subset of CXC chemokines and their receptor CXCR2 are critical to neutrophil recruitment, while CCL3/macrophage inflammatory protein 1α (MIP1α) and CCL2/monocyte chemoattractant protein (MCP)1 are critical to recruitment of monocyte-lineage leukocytes and NK cells, respectively (Phadke & Mehrad 2005). Of activating cytokines, those associated with the Th1 phenotype, including IL12, IL18 and interferon (IFN)γ, are critical for protective responses to the infection. Conversely, the Th2 cytokine IL4 contributes to progression of infection (Mencacci et al 2000, Phadke & Mehrad 2005).

Although epithelial and endothelial cells may internalize conidia (Latge 1999), effector mechanisms of the innate immune system have long been recognized as major host defences against invasive aspergillosis (Latge 1999, Walsh et al 2005). Resident alveolar macrophages ingest inhaled conidia very rapidly, destroy them intracellularly through oxidative mechanisms and prevent germination to hyphae, the invasive form of the fungus. In terminal airways, complement and antibodies cannot be readily available and therefore alveolar macrophages are able to recognize and bind conidia even in the absence of opsonins. In addition, conidia poorly activate the complement system by the classical pathway and, even opsonized, they will trigger only a modest oxidative burst. Polymorphonuclear neutrophils (PMNs), through oxygen and non oxygen-dependent mechanisms, attack hyphae germinating from conidia that escape macrophage surveillance. PMNs are the predominant immune cells in the acute stage of the infection and are essential in initiation and execution of the acute inflammatory response and subsequent resolution of the infection. However, despite extensive fungal growth, pulmonary pathology is reduced in conditions of PMN deficiency both in mice and humans (Marr et al 2002b, Balloy et al 2005), a finding suggesting that PMNs may act as double-edged swords, as the excessive release of oxidants and proteases may be responsible for injury to organs and fungal sepsis. If the above lines of host cellular defences are suppressed (e.g. by corticosteroids) or absent (e.g. as a result of neutropenia), the fungus can germinate into hyphae and invade the lung parenchyma and blood vessels, producing tissue infarction, haemorrhagic necrosis, and death. In some, but not all, patients who remain persistently and profoundly immunocompromised, *A. fumigatus* can disseminate to distal sites including the brain, kidney, liver and skin. The pathogenic determinants responsible for distal seeding of *A. fumigatus* to target organs are unknown.

Dendritic cells

Aspergillus proved to be a useful pathogen model to dissect events occurring at the fungus/dendritic cell (DC) interface. DCs are uniquely able to decode the fungus-associated information and translate it in qualitatively different Th immune responses, *in vitro* and *in vivo* (Grazziutti et al 2001, Bozza et al 2002b, 2003, 2004, Serrano-Gomez et al 2004). By using distinct pattern recognition receptors, including Toll-like receptors (TLRs), human and murine DCs were found to be able to finely discriminate between conidia and hyphae of *Aspergillus* in terms of induction of adaptive Th responses (Bozza et al 2002b, 2003, Bellocchio et al 2004, Romani et al 2004). The fungus has exploited common pathways for entry into DCs, which include a lectin-like pathway for the unicellular form and opsono-dependent pathways for the filamentous form. Recognition and internalization of unopsonized conidia occurred through the engagement of mannose receptors (MRs) of galactomannan specificity, DC-SIGN and, partly, CR3. In contrast, entry of hyphae occurred by a more conventional, zipper-type phagocytosis and involved the cooperative action of FcγR II and III and CR3. Actually, the sugar specificity of MRs involved in the entry of one or multiple *Aspergillus* conidia turned out to be different, as the entry of multiple conidia occurred through a pathway sensitive to galactomannan and that of one single cell through a pathway sensitive to β-glucan (Bozza et al 2002b). Therefore, fungal surface polysaccharides have a key role in the DC/fungus interaction. Transmission electronic microscopy indicated that internalization of conidia occurred predominantly by coiling phagocytosis, characterized by the presence of overlapping bilateral pseudopods that led to a pseudopodal stack before transforming into a phagosome wall. In contrast, entry of hyphae occurred by a more conventional zipper-type phagocytosis, characterized by the presence of symmetrical pseudopods which strictly followed the contour of the hyphae before fusion. However, the fate of the different forms of the fungus inside cells appeared to be quite different. An hour after the exposure, numerous conidia were found inside DCs with no evidence of conidia destruction (Fig. 1), as opposed to hyphae that were rapidly degraded once inside cells. As killing of conidia would seem to be a necessary prerequisite to obtain efficient antigen presentation, it can be postulated that either a small number of conidia are actually

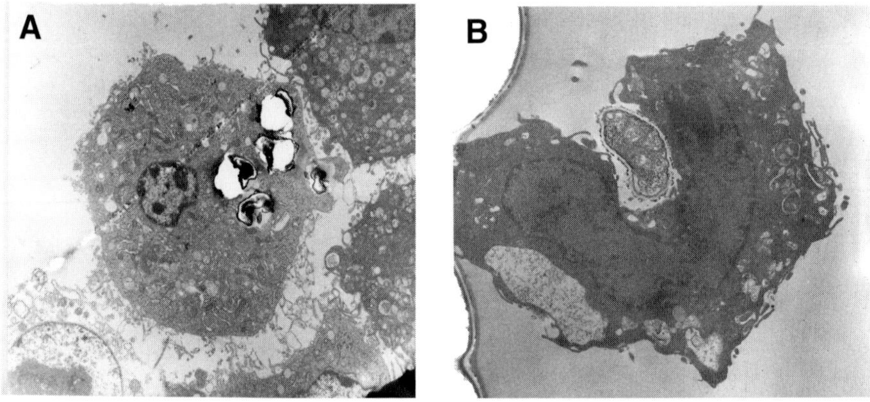

FIG. 1. Transmission electron microscopy of phagocytosis of *Aspergillus fumigatus* by dendritic cells (DCs). Fetal skin derived murine DCs were incubated with live unopsonized *A. fumigatus* conidia (A) or hyphae (B) for 1 h (A) or 3 h (B) before processing for transmission electron microscopy. (A) multiple conidia are seen inside cells (magnification × 20 000) and (B) hyphae uptake through zipper-type phagocytosis (magnification × 8000).

degraded by mature DCs thus allowing their antigen processing and presentation or, alternatively, antigens could be processed and regurgitated by other infected phagocytes and then transferred to DCs for presentation. Entry of *Aspergillus* conidia through MRs resulted in the production of proinflammatory cytokines, including IL12, up-regulation of costimulatory molecules and histocompatibility Class II antigens. IL12 production by DCs required the MyD88 pathway with the implication of distinct TLRs (TLR4 and TLR9). In contrast, coligation of CR3 with FcγR, as in the phagocytosis of hyphae, resulted in the production of IL4/IL10 and up-regulation of costimulatory molecules and histocompatibility Class II antigens. The production of IL10 was largely MyD88-independent (Bellocchio et al 2004). Therefore, TLRs collaborate with other innate immune receptors in the activation of DCs against the fungus through MyD88-dependent and -independent pathways (Fig. 2). It is of interest that TLR gene expression on DCs could be affected upon fungal exposure in a morphotype-dependent manner (Bozza et al 2004) and that the TLR9 agonist CpG-ODN could convert an *Aspergillus* allergen to a potential protective antigen (Bozza et al 2002a) suggesting the potential for TLR agonists to act upon the degree of flexibility of the immune recognition pathways to antigens and allergens. These results suggest that the proper manipulation of DC functioning *in vivo* may translate into beneficial effects in infections. Recent experimental evidence suggests that vaccination against *Aspergillus* through the use of fungus-pulsed DCs is a feasible option (Bozza et al 2003). The infusion of fungus-pulsed or RNA-transfected DCs induced antifungal resistance through the

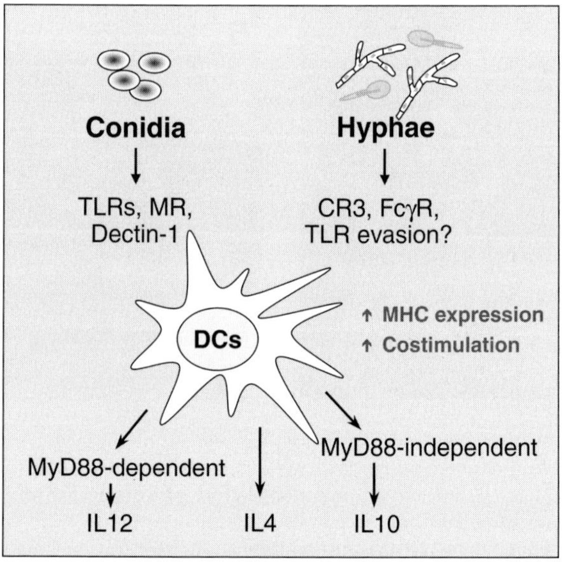

FIG. 2. The exploitation of distinct recognition receptors in dendritic cells (DCs) by *Aspergillus fumigatus* morphotypes. DCs sense fungi in a morphotype-dependent manner. The engagement of distinct receptors on DCs translated into downstream signalling events that differentially affect cytokine production. TLRs, Toll-like receptors; IL1R, IL1 receptor; MR, mannose receptors; CR3, complement receptor 3; FcγR, receptor for the Fc portion of immunoglobulins; MyD88, *Drosophila* myeloid differentiation primary response gene 88.

induction of Th1 cells producing IFNγ. DCs also accelerated the recovery of both myeloid and lymphoid cells in mice with allogenic haematopoietic stem cell transplantation, a finding suggesting that DCs may contribute to the educational program of T cells in haematopoietic stem cell transplantation.

T helper cells

Studies on the epidemiology of invasive aspergillosis (IA) in bone marrow transplantation recipients indicated a reduced neutropenia-related infection and an increased 'late-onset' infection, concomitant with the occurrence of graft-versus-host disease (Marr et al 2002a, Marr et al 2002b). These findings, together with the occurrence in non-neutropenic patients (Denning et al 1991, McCormack & Whitsett 2002), attest to the importance of specific defects in both the innate and adaptive immune effector mechanisms in the pathogenesis of the disease (Roilides et al 1998, Hebart et al 2002). The recent evidence that, in healthy individuals and in patients surviving infection, a significant antigen-specific proliferation of IFNγ-producing T cells occurred (Hebart et al 2002) confirms the crucial role of a Th1 reactivity in the control of infection (Cenci et al 1998, 2000). Two general patterns of Th activation characterize adaptive immune responses in aspergillosis. The Th1 response is associated with increased production of inflammatory cytokines IFNγ, IL2 and IL12 and stimulation of antifungal effector cells (macrophages and PMNs). Alternatively, Th2 responses are associated with suppression of antifungal effector cell activity, decreased production of IFNγ and increased concentrations of IL4 and IL10 which promote humoral responses (IgE) to *Aspergillus* and allergy (Romani 2004). Further evidence indicates that the administration of Th1 type cytokines, such as IFNγ and TNFα, protected mice from a lethal challenge of *Aspergillus*, whereas neutralization of Th2 type cytokines (such as IL4) augmented resistance to the fungus. Conversely, administration of Th2 type cytokines (IL4, IL10) increased susceptibility to the infection and reduced survival (Cenci et al 1998). The importance of Th1/Th2 dysregulation in the outcome of IA in humans was recently supported by work analysing lymphocyte responses in patients with active infection. T cell responses to *A. fumigatus* antigens were compared in healthy patients vs. patients with haematological malignancies who were receiving treatment for probable or proven invasive disease. On exposure to *Aspergillus* antigen, lymphoproliferative responses in healthy individuals exhibited a pattern of increased IFNγ production. Patients with clinical evidence of infectious disease who were responding to antifungal therapy similarly demonstrated strong Th1 lymphoproliferative responses, with IFNγ/IL10 ratios greater than 1.0. Patients with stable or progressive infection on antifungal therapy, however, exhibited poor lymphocyte stimulation indexes and low IFNγ/IL10 ratios consistent with a Th2-predominant response (Hebart et al 2002). Together, clinical and experimental observations suggest that a Th1/Th2 dysregulation with suppression of host Th1 $CD4^+$ lymphocyte response and a switch to a Th2-type immune response may contribute to the development of an unfavourable outcome of aspergillosis.

Regulatory T cells

Th2 cell sensitization to fungal allergens is common in atopic subjects (Kurup 2000), yet respiratory exposure to inhaled conidia is a tolerogenic event in most individuals. It is known that respiratory tolerance is mediated by lung DCs producing IL10 (Akbari et al 2001), which induce the development of $CD4^+$T regulatory cells (T_{reg}) in a costimulation- and TLR-dependent fashion (Mills 2004, O'Garra & Vieira 2004, Belkaid & Rouse 2005). Tolerant T cells express membrane-bound TGFβ and FoxP3 (Ostroukhova et al 2004), the 'master control gene' for the development and function of natural $CD4^+CD25^+$ T_{reg} (Sakaguchi 2005).

Different types of T_{reg} have been defined. Naturally occurring T_{reg} originate in the thymus during the normal process of maturation and survive in the periphery as natural regulators whereas inducible or adaptive T_{reg} develop from conventional $CD4^+$ T cells that are activated in conditions of blockade of costimulatory signals, presence of deactivating cytokines or drugs. $CD4^+CD25^+$ T_{reg} have several modes of suppressive action at their disposal ranging from the inhibitory IL10 and TGFβ to cell–cell contact via the inhibitory CTLA4 (von Boehmer 2005). For the naturally occurring $CD4^+CD25^+$ cells, cell–cell contact-dependent mechanisms have been proposed, while for the adaptive T_{reg} cytokine-dependent mechanisms involving cell-surface TGFβ expression, cell contact-independent mechanisms through soluble IL10 and TGFβ have also been proposed (von Boehmer 2005). Both natural and inducible T_{reg} have been described in infection, their activation occurring through both antigen-specific and non-specific mechanisms (Mills 2004, Belkaid & Rouse 2005). T_{reg} with immunosuppressive activity have been described in fungal infections (Hori et al 2002, Montagnoli et al 2002). By dampening Th1-sterilizing immunity, T_{reg} opposed inflammatory pathology at the expenses of fungal persistence and memory maintenance in candidiasis. Consistent with the notion that basal level expression of B7 costimulatory molecules are required to sustain T_{reg} and CD28/B.7 interactions induce a balance of costimulatory and regulatory signals that have opposite outcomes on immune responses (Lohr et al 2003), the induction of $CD4^+CD25^+$ T_{reg} in candidiasis was strictly dependent on levels of B7 costimulatory antigen expression on IL10-producing Peyer's patches DCs (Montagnoli et al 2002) and involved the IFNγ/IDO-dependent pathway (Bozza et al 2005).

Distinct T_{reg} populations capable of mediating anti-inflammatory or tolerogenic effects were co-ordinately induced by the exposure of mice to *Aspergillus* conidia. Early in infection, a population of T_{reg} expressing the same phenotype as those that control intestinal inflammation and autoimmunity (Alyanakian et al 2003) suppressed lung inflammatory responses to the fungus. Late in infection (and similar to allergy) a population of T_{reg} of the same phenotype as those controlling graft versus host disease (Ermann et al 2005) or diabetes (Alyanakian et al 2003) developed with the ability to control allergic inflammatory response to the fungus. Early in infection, $CD4^+CD25^+$ T cells were particularly increased in the lung of $B7.1^{-/-}$ mice, late in infection in the thoracic lymph nodes of $B7.2^{-/-}$ mice and were neither detected nor increased after infection in $CD28^{-/-}$ or double-deficient $B7.1^{-/-}/B7.2^{-/-}$ mice, a finding pointing to dependency on costimulatory molecule expression. Production of IL10 and TGFβ increased in either type of cells after infection, early $CD25^+$ cells being high producers of IL10 and late $CD25^+$ cells of TGFβ. Therefore, phenotypically distinct $CD4^+CD25^+$ T cell populations are activated in mice exposed to *A. fumigatus*, each population being distinct from Th1 and Th2 effectors and producing IL10 and TGFβ. Consistent with the notion that T_{reg} are capable of directly affecting effector Th cells and inhibiting innate immune cells through inhibitory cytokines (Maloy et al 2003) and IDO (Fallarino et al 2003), late CD25$^+$ T_{reg} inhibited the proliferation of and IL4 production by the corresponding CD25$^-$ population more than early ones did, thus suggesting that Th2 cells were inhibited. In contrast, early T_{reg} suppressed the innate effector functions of PMN and DCs, known to have a central role in the inductive and effector pathways of antifungal immunity (Romani 2004), more than late ones did. Early T_{reg} inhibited the antifungal effector and pro-inflammatory activities of PMN, an activity occurring through both contact-dependent (CTLA4/IDO) and -independent (IL10) mechanisms. These results confirm previous evidence on the occurrence of PMN capable of exhibiting suppressive antifungal effector activity through the CTLA4/IDO-dependent mechanism (Bozza et al 2005) as well as the suppressive activity of IL10 against the fungus (Roilides et al 1997). With respect to DCs, both early and late CD25$^+$ cells inhibited IL12 production of lung DCs in response to conidia but, upon co-culturing fungus-pulsed DCs with late CD25$^+$ cells, high levels of IL10 were also produced, a finding suggesting that, irrespective of the relative contribution of each type of cells to cytokine production, a bidirectional influence may occur between DCs and T_{reg}. Therefore, Th1 cell reactivity was concomitantly down-regulated in the presence of early CD25$^+$ T cells and promoted in the presence of late CD25$^+$ T cells, a finding suggesting that the capacity of early CD25$^+$ T cells (early T_{reg}) to produce anti-inflammatory IL10 sufficient to dampen the inflammatory response to the fungus while the capacity of late CD25$^+$ T cells (late T_{reg}) to produce TGFβ, promoted tolerance to fungal allergy. Selective depletion of early or late T_{reg} exacerbated inflammation and allergy to the fungus, whereas adoptive transfer of early or late T_{reg} restored resistance to both infection and allergy, both findings indicating a causal link between resistance/susceptibility to infection and allergy and the activity of the distinct T_{reg} subsets.

T_{reg} induction and function were strictly dependent on the IFNγ/IDO-dependent axis acting on both PMNs and DCs. Not only was IDO functional activity positively correlated with the suppressive activity of early T_{reg} on PMNs, but IDO blockade exacerbated allergy, a finding that suggests loss of tolerogenic T_{reg}. Therefore, IDO serves a crucial role in *Aspergillus* infection and allergy and is involved in both T_{reg} functioning and induction. Consistent with the finding that IFNγ is one major activating signal for IDO (Fallarino et al 2003), the impaired IDO expression and functional activity observed in conditions of IFNγ deficiency was concomitant with defective functioning of early T_{reg} and defective occurrence of late T_{reg}. Therefore, low levels of IFNγ production early in infection are associated with defective activation of tolerogenic T_{reg}, which links inflammatory events occurring at the early stages of the infection to subsequent allergic responses to the fungus through IFNγ/IDO.

Conclusions

Regulatory mechanisms operating in the control of inflammation and allergy to fungi are different but interdependent as the level of the inflammatory response early in infection may impact on susceptibility to allergy, in conditions of continuous exposure to the fungus. IDO has a unique and central role in this process as it participates in the effector and inductive phases of early and tolerogenic T_{reg}. This may explain the beneficial effect on fungal allergy of CpG oligodeoxynucleotides (Banerjee et al 2004) known to induce IDO that is found to inhibit experimental asthma (Hayashi et al 2004) and to have increased activity in asymptomatic atopy (von Bubnoff et al 2004). The data are compatible with a scenario in which a division of labour occurs between functionally distinct T_{reg} populations that are co-ordinately activated upon the exposure to *Aspergillus* (Fig. 3). Early on in infection, antiinflammatory T_{reg}, requiring B7.2/CD28 for generation, suppress through the combined action of IL10 and CTLA4 acting on IDO. The concurrent engagement of T_{reg} and effector cells at inflammatory sites allows immune responses to be vigorous enough to provide adequate host defense, without necessarily eliminat-

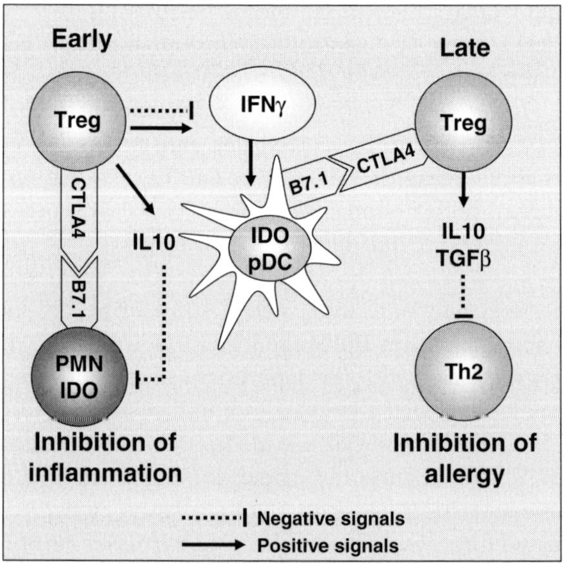

FIG. 3. The central role of the IFNγ/IDO-dependent pathway in immunity and tolerance to *Aspergillus* and its subversion by the fungus. The production of IFNγ is squarely placed at the host/pathogen interface where IDO activation exerts a fine control over the inductive and effector pathways of immunity and tolerance to *A. fumigatus* infection and allergy. pDC, plasmacytoid DC.

ing the fungus or causing an unacceptable level of host damage. Indeed, the levels of IFNγ produced in this early phase set the subsequent adaptive stage by conditioning the tolerogenic program of DCs through IDO. This results in the occurrence of tolerogenic T_{reg} producing IL10 and TGFβ and inhibiting allergic Th2 cells. As an association has been reported between serum IL10 levels and the progression of invasive aspergillosis in non-neutropenic patients (Roilides et al 2001), it is tempting to speculate that, irrespective of the underlying immunosuppressive disease state, a dysregulated T_{reg} functioning may both predispose and be a surrogate marker for identifying patients at risk for *Aspergillus* infection.

Acknowledgements

We thank Lara Bellocchio for dedicated editorial assistance. This study was supported by the National Research Project on AIDS, contract 50F.30, 'Opportunistic Infections and Tuberculosis', Italy.

References

Akbari O, DeKruyff RH, Umetsu DT 2001 Pulmonary dendritic cells producing IL-10 mediate tolerance induced by respiratory exposure to antigen. Nat Immunol 2:725–731

Alyanakian MA, You S, Damotte D et al 2003 Diversity of regulatory CD4+T cells controlling distinct organ-specific autoimmune diseases. Proc Natl Acad Sci USA 100:15806–15811

Balloy V, Huerre M, Latge JP, Chignard M 2005 Differences in patterns of infection and inflammation for corticosteroid treatment and chemotherapy in experimental invasive pulmonary aspergillosis. Infect Immun 73:494–503

Banerjee B, Kelly KJ, Fink JN, Henderson JD, Bansal NK, Kurup VP 2004 Modulation of airway inflammation by immunostimulatory CpG oligodeoxynucleotides in a murine model of allergic aspergillosis. Infect Immun 72:6087–6094

Belkaid Y, Rouse BT 2005 Natural regulatory T cells in infectious disease. Nat Immunol 6:353–360

Bellocchio S, Montagnoli C, Bozza S et al 2004 The contribution of the toll-like/IL-1 receptor superfamily to innate and adaptive immunity to fungal pathogens in vivo. J Immunol 172:3059–3069

Bozza S, Gaziano R, Lipford GB et al 2002a Vaccination of mice against invasive aspergillosis with recombinant Aspergillus proteins and CpG oligodeoxynucleotides as adjuvants. Microbes Infect 4:1281–1290

Bozza S, Gaziano R, Spreca A et al 2002b Dendritic cells transport conidia and hyphae of Aspergillus fumigatus from the airways to the draining lymph nodes and initiate disparate Th responses to the fungus. J Immunol 168:1362–1371

Bozza S, Perruccio K, Montagnoli C et al 2003 A dendritic cell vaccine against invasive aspergillosis in allogeneic hematopoietic transplantation. Blood 102:3807–3814

Bozza S, Montagnoli C, Gaziano R et al 2004 Dendritic cell-based vaccination against opportunistic fungi. Vaccine 22:857–864

Bozza S, Fallarino F, Pitzurra L et al 2005 A crucial role for tryptophan catabolism at the host/Candida albicans interface. J Immunol 174:2910–2918

Cenci E, Mencacci A, Fe d'Ostiani C et al 1998 Cytokine- and T helper-dependent lung mucosal immunity in mice with invasive pulmonary aspergillosis. J Infect Dis 178:1750–1760

Cenci E, Mencacci A, Bacci A, Bistoni F, Kurup VP, Romani L 2000 T cell vaccination in mice with invasive pulmonary aspergillosis. J Immunol 165:381–388

Cenci E, Mencacci A, Casagrande A, Mosci P, Bistoni F, Romani L 2001 Impaired antifungal effector activity but not inflammatory cell recruitment in interleukin-6-deficient mice with invasive pulmonary aspergillosis. J Infect Dis 184:610–617

Denning DW, Follansbee SE, Scolaro M, Norris S, Edelstein H, Stevens DA 1991 Pulmonary aspergillosis in the acquired immunodeficiency syndrome. N Engl J Med 324:654–662

Ermann J, Hoffmann P, Edinger M et al 2005 Only the CD62L+ subpopulation of CD4+CD25+ regulatory T cells protects from lethal acute GVHD. Blood 105:2220–2226

Fallarino F, Grohmann U, Hwang KW et al 2003 Modulation of tryptophan catabolism by regulatory T cells. Nat Immunol 4:1206–1212

Garlanda C, Hirsch E, Bozza S et al 2002 Non-redundant role of the long pentraxin PTX3 in anti-fungal innate immune response. Nature 420:182–186

Grazziutti M, Przepiorka D, Rex JH, Braunschweig I, Vadhan-Raj S, Savary CA 2001 Dendritic cell-mediated stimulation of the in vitro lymphocyte response to aspergillus. Bone Marrow Transplant 27:647–652

Hayashi T, Beck L, Rossetto C et al 2004 Inhibition of experimental asthma by indoleamine 2,3-dioxygenase. J Clin Invest 114:270–279

Hebart H, Bollinger C, Fisch P et al 2002 Analysis of T-cell responses to Aspergillus fumigatus antigens in healthy individuals and patients with hematologic malignancies. Blood 100:4521–4528

Hori S, Carvalho TL, Demengeot J 2002 CD25+CD4+ regulatory T cells suppress CD4+ T cell-mediated pulmonary hyperinflammation driven by pneumocystis carinii in immunodeficient mice. Eur J Immunol 32:1282–1291

Kurup VP 2000 Immunology of allergic bronchopulmonary aspergillosis. Indian J Chest Dis Allied Sci 42:225–237

Latge J P 1999 Aspergillus fumigatus and aspergillosis. Clin Microbiol Rev 12:310–350

Lohr J, Knoechel B, Jiang S, Sharpe AH, Abbas AK 2003 The inhibitory function of B7 costimulators in T cell responses to foreign and self-antigens. Nat Immunol 4:664–669

Madan T, Kaur S, Saxena S et al 2005 Role of collectins in innate immunity against aspergillosis. Med Mycol 43:S155–163

Maloy KJ, Salaun L, Cahill R, Dougan G, Saunders NJ, Powrie F 2003 CD4+CD25+ T(R) cells suppress innate immune pathology through cytokine-dependent mechanisms. J Exp Med 197:111–119

Marr KA, Carter RA, Boeckh M, Martin P, Corey L 2002a Invasive aspergillosis in allogeneic stem cell transplant recipients: changes in epidemiology and risk factors. Blood 100:4358–4366

Marr KA, Patterson T, Denning D 2002b Aspergillosis. Pathogenesis, clinical manifestations, and therapy. Infect Dis Clin North Am 16:875–894

McCormack FX, Whitsett JA 2002 The pulmonary collectins, SP-A and SP-D, orchestrate innate immunity in the lung. J Clin Invest 109:707–712

Mencacci A, Cenci E, Bacci A, Montagnoli C, Bistoni F, Romani L 2000 Cytokines in candidiasis and aspergillosis. Curr Pharm Biotechnol 1:235–251

Mills KH 2004 Regulatory T cells: friend or foe in immunity to infection? Nat Rev Immunol 4:841–855

Montagnoli C, Bacci A, Bozza S et al 2002 B7/CD28-dependent CD4+CD25+ regulatory T cells are essential components of the memory-protective immunity to Candida albicans. J Immunol 169:6298–6308

O'Garra A, Vieira P 2004 Regulatory T cells and mechanisms of immune system control. Nat Med 10:801–805

Ostroukhova M, Seguin-Devaux C, Orris TB et al 2004 Tolerance induced by inhaled antigen involves CD4(+) T cells expressing membrane-bound TGF-beta and FOXP3. J Clin Invest 114:28–38

Phadke AP, Mehrad B 2005 Cytokines in host defense against Aspergillus: recent advances. Med Mycol 43:S173–176

Roilides E, Dimitriadou A, Kadiltsoglou I et al 1997 IL-10 exerts suppressive and enhancing effects on antifungal activity of mononuclear phagocytes against Aspergillus fumigatus. J Immunol 158:322–329

Roilides E, Katsifa H, Walsh TJ 1998 Pulmonary host defences against Aspergillus fumigatus. Res Immunol 149:454–465

Roilides E, Sein T, Roden M, Schaufele RL, Walsh TJ 2001 Elevated serum concentrations of interleukin-10 in nonneutropenic patients with invasive aspergillosis. J Infect Dis 183:518–520

Romani L 2004 Immunity to fungal infections. Nat Rev Immunol 4:1–23

Romani L, Bistoni F, Gaziano R et al 2004 Thymosin alpha 1 activates dendritic cells for antifungal Th1 resistance through toll-like receptor signaling. Blood 103:4232–4239

Sakaguchi S 2005 Naturally arising Foxp3-expressing CD25+CD4+ regulatory T cells in immunological tolerance to self and non-self. Nat Immunol 6:345–352

Serrano-Gomez D, Dominguez-Soto A, Ancochea J, Jimenez-Heffeman JA, Leal JA, Corbi AL 2004 Dendritic cell-specific intercellular adhesion molecule 3-grabbing nonintegrin mediates binding and internalization of Aspergillus fumigatus conidia by dendritic cells and macrophages. J Immunol 173:5635–5643

von Boehmer H 2005 Mechanisms of suppression by suppressor T cells. Nat Immunol 6:338–344

von Bubnoff D, Fimmers R, Bogdanow M, Matz H, Koch S, Bieber T 2004 Asymptomatic atopy is associated with increased indoleamine 2,3-dioxygenase activity and interleukin-10 production during seasonal allergen exposure. Clin Exp Allergy 34:1056–1063

Walsh TJ, Roilides E, Cortez H, Kottillil S, Bailey J, Lyman CA 2005 Control, immunoregulation, and expression of innate pulmonary host defenses against Aspergillus fumigatus. Med Mycol 43:S165–172

DISCUSSION

Finn: When you talk about the development of T_{reg} and even effector cells, you were suggesting that there can be extra-lymph node development of these cells. Then you said that they have homing molecules to go to the lymph node and do whatever they need to do. Is this a well accepted process? Do we actually generate T cell responses outside the setting of the lymph node?

Romani: What I think is happening is that a pool of T_{reg} exists that recirculates into the lung where they can get activated. As a matter of fact, resting conidia may recruit T_{reg}. Some of them will undergo maturation in the sense that they will acquire the CCR7 marker, which lets them go to the thoracic lymph node where they will act upon Th cells.

Latgé: You mentioned that there was some specific interaction between *A. fumigatus* and the lung in the host reaction. Do you think this is because we have only looked at this pathogen? I'm sure that any thermophilic fungus that gets into your lung will destroy the lung tissues and induce an immune response that will be similar to the one seen with *A. fumigatus*. For example, if you take a strawberry saphrophyte like *Neosartorya fischeri* that can grow at 37 °C and put it into the lung of a mouse, the mouse will die. Do you think there is something common to all thermophilic fungi that could be explaining this type of reaction?

Romani: It is possible. The T_{reg} story indicates to us that there is an inherent capacity of the lung to regulate the inflammatory response to any pathogen.

Latgé: For TIM, you said that *Aspergillus* is binding to pentraxin 3 but not *Candida*. Do you know which molecule is involved?

Romani: Yes, it is galactomannan. We have found that galactomannan from *Aspergillus* is a major ligand.

Latgé: The SIGMA galactomannan that has been used is not specific to *Aspergillus* and has indeed a composition that is chemically very different from the *Aspergillus* galactomannan.

Romani: We have looked for something that was present on *Aspergillus* but not in *Candida* and that was recognized by pentraxin. We have inhibited the binding with galactomannan.

Brown: Could you comment on the inhibition of inflammation?

Romani: By induction of indoleamine 2, 3-dioxygenase (IDO) in dendritic cells (DCs) and phagocytes. Up-regulation of IDO results in down-regulation of the inflammatory response. For instance, upon IDO blocking, neutrophils are no longer able to produce TNFα. In DCs, the balance between IL12 and IL10 is skewed towards IL10 in the condition of IDO blockade, which means that these DCs are able to prime T_{reg}.

Lambrecht: In your last scheme you came up with plasmacytoid dendritic cells (pDCs) as being tolerogenic cells for these so-called late T_{reg}. We found something very similar in response to harmless ovalbumin inhalation. When we deplete our pDCs we can break inhalation tolerance. In your system, what is the evidence for pDCs being tolerogenic cells? Do they present the antigen? We had a hard time proving that they can actually present inhaled antigens. Or is this just a generalized attraction of T_{reg} induced by pDCs?

Romani: There are different types of pDCs. The pDCs we grow from mouse bone marrow and human peripheral blood do present *Aspergillus* antigens, and activate either Th2 responses or T_{reg} responses depending on the level of IL10 being produced. This is something that we have already found, and more than this, we have transferred these pDCs, as well as myeloid DCs, into immunocompetent mice and found that myeloid DCs induced an inflammatory response which is not associated with protection from aspergillosis. Actually, the mice died earlier. If we inject pDCs, by activation of both Th1 and T_{reg} responses, we restored resistance in otherwise susceptible mice.

Latgé: I was interested in the Toll-like receptor (TLR) story. In the patients, have you seen any effect on TLR for those who are going to get *Aspergillus* or those who resist infection?

Romani: No, because there is no way to measure this.

E Sim: I was interested in what you said about pentraxin 3. My colleagues in Oxford have done interesting experiments with another soluble recognition molecule, the lung collectin SPE. They showed that if you take SPE knockout mice, they are susceptible to *A. fumigatus.* If they add back a recombinant form of SPE it gives resistance to both infection and allergy. The particular form of SPE they use is a truncated form that lacks the collagen region, which is the region that might be expected to interact with the receptor. The mechanism by which the recombinant form works is not very clear. Do you have any more information in terms of the pentraxin 3? Is it acting directly as an opsonin?

Romani: It works like an opsonin. Moreover, we have also found that pentraxin 3 binds and has a trophic effect on DCs from the lung. Somehow pentraxin 3 is acting as a soluble or pattern recognition receptor that influences the survival state of the cells. DCs from haematopoietic transplanted patients recovered nicely if flooded with pentraxin 3.

Pentraxins in innate immunity and inflammation

Cecilia Garlanda, Barbara Bottazzi, Giovanni Salvatori*, Rita De Santis*, Alessia Cotena, Livija Deban, Viriginia Maina, Federica Moalli, Andrea Doni, Tania Veliz-Rodriguez and Alberto Mantovani[1†]

*Istituto Clinico Humanitas, Via Manzoni, 56, 20089 Rozzano (Milan), Italy *Immunology Area, R&D Department, Sigma-Tau Industrie Farmaceutiche Riunite SpA, via Pontina, 00040 Pomezia, Rome, Italy and † University of Milan, Italy*

Abstract. C-reactive protein, the first innate immunity receptor identified, and serum amyloid P component are classic short pentraxins produced in the liver. Long pentraxins, the prototype of which is PTX3, are expressed in a variety of tissues. PTX3 is produced by a variety of cells and tissues, most notably dendritic cells and macrophages, in response to TLR engagement and inflammatory cytokines. PTX3 acts as a functional ancestor of antibodies, recognizing microbes, activating complement, facilitating pathogen recognition by phagocytes, hence playing a non-redundant role in resistance against selected pathogens, in particular in the lung. Thus, the prototypic long pentraxin PTX3 is a multifunctional soluble pattern recognition receptor at the crossroads between innate immunity, inflammation, matrix deposition and female fertility.

2006 Innate immunity to pulmonary infection. Wiley, Chichester (Novartis Foundation Symposium 279) p 80–91

The classic short pentraxins C-reactive protein (CRP) and serum amyloid P component (SAP) are acute phase proteins in human and mouse, respectively. The liver produces these molecules in response to inflammatory signals, most prominently interleukin (IL)6. The prototypic long pentraxin 3 (PTX3) has similarities with the classical short pentraxins. However, it has an unrelated long N-terminal domain coupled to the C-terminal pentraxin domain, and differs in gene organization, cellular source and ligands recognized (Garlanda et al 2005). Several cell types rapidly produce and release PTX3 (mononuclear phagocytes, dendritic cells [DCs], fibroblasts and endothelial cells) (Doni et al 2003, Garlanda et al 2005), in response to Toll-like receptor (TLR) engagement, tumour necrosis factor (TNF) α and IL1β.

[1]This paper was presented at the symposium by Alberto Mantovani, to whom correspondence should be addressed.

PTX3 binds with high affinity the complement component C1q, the extracellular matrix component TNFα-induced protein 6 (TNFAIP6 or TSG-6) and selected microorganisms, including *Aspergillus fumigatus* and *Pseudomonas aeruginosa* (Bottazzi et al 1997, Garlanda et al 2002, Nauta et al 2003, Diniz et al 2004, Salustri et al 2004). PTX3 activates the classical pathway of complement activation and facilitates pathogen recognition by macrophages and DCs (Garlanda et al 2002, 2005, Diniz et al 2004).

PTX3 plays complex non-redundant functions *in vivo*, ranging from the assembly of a hyaluronic acid-rich extracellular matrix and female fertility, to innate immunity against selected microbial agents (Dias et al 2001, Garlanda et al 2002, 2005, Souza et al 2002, Diniz et al 2004, Salustri et al 2004). PTX3 is highly conserved in evolution. Evidence suggests that PTX3 is an important component of the humoral arm of innate immunity, downstream of, and complementary to, cellular recognition and activation, with a function in recognition and defence against pulmonary pathogens.

Production of PTX3

PTX3 is produced by a variety of cell types upon exposure to IL1β, TNFα, microbial moieties such as lipopolysaccharide (LPS), lipoarabinomannans, Outer membrane protein A (OmpA) (Jeannin et al 2005) and agonists for different TLRs. These cells include myeloid DC, that are major producers of PTX3, endothelial cells, mononuclear phagocytes, smooth muscle cells, adipocytes, fibroblasts, synovial cells and chondrocytes (Goodman et al 2000, Doni et al 2003, Klouche et al 2004). Recently, cells of epithelial origin, in particular alveolar epithelial cells, have also been found to produce low amounts of PTX3 under stimulation (dos Santos et al 2004). IL6, a poor inducer of PTX3 *in vitro*, was found to be involved in PTX3 expression in Castleman's disease (Malaguarnera et al 2000) and in Kaposi's sarcoma (Klouche et al 2002).

IFNγ and IL10 have divergent effects on PTX3 production. IFNγ, which has generally a synergistic effect with LPS (Ehrt et al 2001), inhibits LPS-induced PTX3 expression and production in different cellular contexts (Goodman et al 2000), whereas IL10 weakly induces PTX3 expression in DC and monocytes and significantly synergizes with LPS, other TLR agonists and IL1β (Perrier et al 2004, Doni et al 2006). IL10 induces a set of genes (e.g. type I collagen, fibronectin, versican, α1-antitrypsin) related to tissue remodelling (Lang et al 2002, Perrier et al 2004) and is involved in the chronic and resolution phase of inflammation. Given its role in matrix organization (Salustri et al 2004), PTX3 expression in M2 mononuclear phagocytes and IL10-treated DCs and fibroblasts is likely to be related to the orchestration of matrix deposition, tissue repair and remodelling. Moreover, it is interesting that beside the stimulation of B cell differentiation and antibody production, the humoral arm of adaptive (antibodies) immunity, IL10 stimulates also the humoral arm of innate (PTX3) immunity.

Ligand recognition and effector functions

PTX3 binds different ligands including the complement component C1q (Bottazzi et al 1997, Nauta et al 2003), the growth factor fibroblast growth factor 2 (FGF2) (Rusnati et al 2004), the extracellular matrix protein TSG-6 (Salustri et al 2004) and the outer membrane protein A from *Klebsiella pneumoniae* (KpOmpA) (Jeannin et al 2005). As classical short pentraxins do, PTX3 binds to plastic-immobilized C1q inducing complement activation (Bottazzi et al 1997, Nauta et al 2003). In contrast, fluid-phase binding of PTX3 to C1q inhibits complement activation by blocking relevant interaction sites (Nauta et al 2003). While interaction of PTX3 with C1q is calcium independent, the presence of calcium is required for PTX3 interaction with other ligands, such as TSG-6 and KpOmpA (Bottazzi et al 1997, Jeannin et al 2005 and Bottazzi B, unpublished data). In addition PTX3 enhances the deposition of both C1q and C3 on apoptotic cells (Nauta et al 2003). These data further support accumulating evidence suggesting that complement components and pentraxins may participate in the handling of apoptotic cells (Nauta et al 2003).

PTX3 deficient mice are more susceptible to invasive pulmonary aspergillosis than control wild type animals. This can be explained, at least in part, by an opsonic effect of PTX3 facilitating ingestion of conidia by macrophages (Garlanda et al 2002). Macrophages from PTX3-transgenic mice have an improved phagocytic activity towards zymosan and *Paracoccidioides brasiliensis* (Diniz et al 2004). Moreover recombinant PTX3 binds to zymosan and *P. brasiliensis* and functions as an opsonin increasing the phagocytic activity of peritoneal macrophages from wild-type animals. These findings provide evidence for a role of PTX3 as an opsonin and imply the existence of a receptor for this molecule. PTX3 in fact binds in a dose-dependent and saturable way murine macrophages as well as human mononuclear phagocytes and DC (B. Bottazzi, unpublished data).

Innate resistance and inflammation

PTX3 is non-redundant in selected fungal and bacterial infections (*A. fumigatus*, *P. aeruginosa*, *S. typhymurium*) and irrelevant in others (*L. monocytogenes*, *S. aureus*, polymicrobic intra-abdominal sepsis) (Garlanda et al 2002, and C. Garlanda, unpublished data). PTX3 deficiency does not cause a generalized impairment of host resistance to microbial pathogens, and PTX3 is involved in recognition and resistance against specific microorganisms. In particular, PTX3 deficient mice were extremely susceptible to invasive pulmonary aspergillosis and the specificity of the defect and the therapeutic potential of PTX3 could be demonstrated by the complete protective effect of the treatment with recombinant PTX3 (Garlanda et al 2002, Gaziano et al 2004). Moreover, in this model the defective recognition of *A. fumigatus* conidia by PTX3-deficient mice was associated with the lack of development of appropriate and protective T helper-cell type 1 (Th1) anti-fungal responses and to an unbalanced cytokine profile skewed towards a Th2 response (Garlanda et al 2002).

PTX3 binds OmpA from *Klebsiella pneumoniae* (KpOmpA). Cellular recognition of KpOmpA, a major conserved outer membrane protein of Gram-negative *enterobacteriaceae*, is mediated by two members of the SR family, LOX1 and SRECI, expressed on macrophages and DC, whereas cellular activation by KpOmpA is cooperatively mediated by TLR2 (Jeannin et al 2000, Massari et al 2002). Activation of cellular innate immunity by KpOmpA is followed by induction of PTX3, a component of the humoral arm of innate immunity, which in turns binds KpOmpA with high affinity (Jeannin et al 2005). Defective local inflammation elicited by KpOmpA observed in TLR2 and PTX3 deficient mice supports that both the cellular and the humoral arms of innate immunity are essential for a full response to KpOmpA. Moreover, in a model of local inflammation, PTX3 significantly amplifies the inflammatory response elicited by OmpA acting as a non-redundant humoral amplification system of the response elicited by KpOmpA. Thus, cellular and humoral recognition of KpOmpA are complementary in mediating the innate response to this conserved key component of *enterobacteriaceae.*

PTX3 behaves as an acute phase response protein since its blood levels, low in normal conditions (about 25 ng/ml in the mouse, <2 ng/ml in human), increase rapidly (peak at 6–8 h) and dramatically (200–800 ng/ml) during endotoxic shock, sepsis and other inflammatory and infectious conditions correlating with the severity of the disease (Luchetti et al 2000, Fazzini et al 2001, Muller et al 2001, Latini et al 2004, Azzurri et al 2005, Mairuhu et al 2005). In these conditions, PTX3 could serve as a marker for primary local activation of innate immunity and inflammation. Indeed, in all clinical studies conducted so far, the correlation between levels of PTX3 and CRP was loose or non-significant (Muller et al 2001, Latini et al 2004). It remains to be elucidated whether the impressive correlation with outcome and severity actually reflects a role in the pathogenesis of damage, for instance by amplifying the complement and coagulation cascades (Napoleone et al 2002, 2004).

The *in vivo* role of PTX3 in inflammatory conditions has been investigated using PTX3 overexpressing and deficient mice. In a model of LPS toxicity and in cecal ligation and puncture, PTX3 overexpression resulted in increased resistance (Dias et al 2001), whereas its deficiency was irrelevant (Garlanda et al 2002). Following intestinal ischaemia reperfusion injury, PTX3 overexpressing mice showed exacerbated inflammatory response and reduced survival rate, due to enhanced production of pro-inflammatory mediators (TNFα in particular) (Souza et al 2002).

In conclusion, PTX3, produced by DCs, neighbouring macrophages and other cell types upon TLRs engagement or pathogen recognition, recognizes microbial moieties, opsonizes fungi and selected Gram positive and Gram negative bacteria and activates complement. Opsonization results in facilitated pathogen recognition (increased phagocytosis and killing), innate immune cell activation (increased cytokine and nitric oxide production); moreover, opsonization by PTX3 is likely to be involved in the activation of an appropriate adaptive immune response (DC maturation and polarization). Thus, this long pentraxin behaves as a *bona fide* predecessor of antibodies.

Pathogen versus apoptotic self discrimination

PTX3 binds late apoptotic cells inhibiting their recognition by DC (Rovere et al 2000). In addition, preincubation of apoptotic cells with PTX3 enhances C1q binding and C3 deposition on the cell surface, suggesting a role for PTX3 in the complement-mediated clearance of apoptotic cells (Nauta et al 2003).

PTX3 can regulate secretion of IL10 and TNFα by DCs, suggesting that PTX3 behaves as a flexible regulator of the functions of this cell population (Baruah et al 2005). Moreover, in the presence of dying cells, PTX3 restricts the cross presentation of antigens derived from dying cells (Baruah et al 2005). These results have led to the speculation that PTX3 has a dual role in the protection against pathogens and in the control of autoimmunity.

Concluding remarks

CRP was the first pattern recognition receptor to be identified (Garlanda et al 2005). However, its *in vivo* function has not been unequivocally defined. Indeed, the considerable differences in sequence and, most prominently, regulation (CRP is not an acute-phase protein in mouse) have precluded the use of straightforward genetic approaches to explore its *in vivo* function. In contrast, gene targeting of the prototypic, evolutionary conserved, long pentraxin PTX3 has unequivocally defined the role of this molecule and, by inference, of the whole pentraxins family, at the crossroad of innate immunity, inflammation, matrix deposition and female fertility (Garlanda et al 2005). In particular, gene targeted mice have revealed that PTX3 and, by influence, other pentraxins are essential for lung resistance against fungal and bacterial pathogens.

Acknowledgements

This work was supported by Associazione Italiana per la Ricerca sul Cancro (AIRC), Telethon, Ministero Istruzione Università e Ricerca (MIUR), Ministero della Salute, CNR, European Commission.

References

Azzurri A, Sow OY, Amedei A et al 2005 IFN-γ-inducible protein 10 and pentraxin 3 plasma levels are tools for monitoring inflammation and disease activity in Mycobacterium tuberculosis infection. Microbes Infect 7:1–8

Baruah P, Propato A, Dumitriu IE et al 2005 The pattern recognition receptor PTX3 is recruited at the synapse between dying and dendritic cells and edits the cross-presentation of self, viral and tumor antigens. Blood 107:151–158

Bottazzi B, Vouret-Craviari V, Bastone A et al 1997 Multimer formation and ligand recognition by the long pentraxin PTX3. Similarities and differences with the short pentraxins C-reactive protein and serum amyloid P component. J Biol Chem 272:32817–32823

Dias AA, Goodman AR, Dos Santos JL et al 2001 TSG-14 transgenic mice have improved survival to endotoxemia and to CLP-induced sepsis. J Leukoc Biol 69:928–936

Diniz SN, Nomizo R, Cisalpino PS et al 2004 PTX3 function as an opsonin for the dectin-1-dependent internalization of zymosan by macrophages. J Leukoc Biol 75:649–656

Doni A, Mosca M, Bottazzi B et al 2006 Regulation of PTX3, a key component of humoral innate immunity, in human dendritic cells: stimulation by IL-10 and inhibition by IFN-gamma. J Leukoc Biol 79:797–802

Doni A, Peri G, Chieppa M et al 2003 Production of the soluble pattern recognition receptor PTX3 by myeloid, but not plasmacytoid, dendritic cells. Eur J Immunol 33:2886–2893

dos Santos CC, Han B, Andrade CF et al 2004 DNA microarray analysis of gene expression in alveolar epithelial cells in response to TNFα, LPS, and cyclic stretch. Physiol Genomics 19:331–342

Ehrt S, Schnappinger D, Bekiranov S et al 2001 Reprogramming of the macrophage transcriptome in response to interferon-gamma and Mycobacterium tuberculosis: signaling roles of nitric oxide synthase-2 and phagocyte oxidase. J Exp Med 194:1123–1140

Fazzini F, Peri G, Doni A et al 2001 PTX3 in small-vessel vasculitides: an independent indicator of disease activity produced at sites of inflammation. Arthritis Rheum 44:2841–2850

Garlanda C, Bottazzi B, Bastone A et al 2005 Pentraxins at the crossroads between innate immunity, inflammation, matrix deposition, and female fertility. Annu Rev Immunol 23:337–366

Garlanda C, Hirsch E, Bozza S et al 2002 Non-redundant role of the long pentraxin PTX3 in anti-fungal innate immune response. Nature 420:182–186

Gaziano R, Bozza S, Bellocchio S et al 2004 Anti-Aspergillus fumigatus efficacy of pentraxin 3 alone and in combination with antifungals. Antimicrob Agents Chemother 48:4414–4421

Goodman AR, Levy DE, Reis LF et al 2000 Differential regulation of TSG-14 expression in murine fibroblasts and peritoneal macrophages. J Leukoc Biol 67:387–395

Jeannin P, Bottazzi B, Sironi M et al 2005 Complexity and complementarity of outer membrane protein A recognition by cellular and humoral innate immunity receptors. Immunity 22:551–560

Jeannin P, Renno T, Goetsch L et al 2000 OmpA targets dendritic cells, induces their maturation and delivers antigen into the MHC class I presentation pathway. Nat Immunol 1:502–509

Klouche M, Brockmeyer N, Knabbe C et al 2002 Human herpesvirus 8-derived viral IL-6 induces PTX3 expression in Kaposi's sarcoma cells. Aids 16:F9–18

Klouche M, Peri G, Knabbe C et al 2004 Modified atherogenic lipoproteins induce expression of pentraxin-3 by human vascular smooth muscle cells. Atherosclerosis 175:221–228

Lang R, Patel D, Morris JJ et al 2002 Shaping gene expression in activated and resting primary macrophages by IL-10. J Immunol 169:2253–2263

Latini R, Maggioni AP, Peri G et al 2004 Prognostic significance of the long pentraxin PTX3 in acute myocardial infarction. Circulation 110:2349–2354
Luchetti MM, Piccinini G, Mantovani A et al 2000 Expression and production of the long pentraxin PTX3 in rheumatoid arthritis (RA). Clin Exp Immunol 119:196–202
Mairuhu AT, Peri G, Setiati TE et al 2005 Elevated plasma levels of the long pentraxin, pentraxin 3, in severe dengue virus infections. J Med Virol 76:547–552
Malaguarnera L, Pilastro MR, Vicari L et al 2000 PTX3 gene expression in Castleman's disease. Eur J Haematol 64:132–134
Massari P, Henneke P, Ho Y et al 2002 Cutting edge: Immune stimulation by neisserial porins is toll-like receptor 2 and MyD88 dependent. J Immunol 168:1533–1537
Muller B, Peri G, Doni A et al 2001 Circulating levels of the long pentraxin PTX3 correlate with severity of infection in critically ill patients. Crit Care Med 29:1404–1407
Napoleone E, Di Santo A, Bastone A et al 2002 Long pentraxin PTX3 upregulates tissue factor expression in human endothelial cells: a novel link between vascular inflammation and clotting activation. Arterioscler Thromb Vasc Biol 22:782–787
Napoleone E, di Santo A, Peri G et al 2004 The long pentraxin PTX3 up-regulates tissue factor in activated monocytes: another link between inflammation and clotting activation. J Leukoc Biol 76:203–209
Nauta AJ, Daha MR, van Kooten C et al 2003 Recognition and clearance of apoptotic cells: a role for complement and pentraxins. Trends Immunol 24:148–154
Perrier P, Martinez FO, Locati M et al 2004 Distinct transcriptional programs activated by interleukin-10 with or without lipopolysaccharide in dendritic cells: induction of the B cell-activating chemokine, CXC chemokine ligand 13. J Immunol 172:7031–7042
Rovere P, Peri G, Fazzini F et al 2000 The long pentraxin PTX3 binds to apoptotic cells and regulates their clearance by antigen-presenting dendritic cells. Blood 96:4300–4306
Rusnati M, Camozzi M, Moroni E et al 2004 Selective recognition of fibroblast growth factor-2 by the long pentraxin PTX3 inhibits angiogenesis. Blood 104:92–99
Salustri A, Garlanda C, Hirsch E et al 2004 PTX3 plays a key role in the organization of the cumulus oophorus extracellular matrix and in *in vivo* fertilization. Development 131:1577–1586
Souza DG, Soares AC, Pinho V et al 2002 Increased mortality and inflammation in tumor necrosis factor-stimulated gene-14 transgenic mice after ischemia and reperfusion injury. Am J Pathol 160:1755–1765

DISCUSSION

Ryffel: It is fascinating that with the short and long term pentraxins you can make a distinction between systemic and local. The systemic ones, such as C-reactive protein (CRP), are also produced in macrophages or innate immune cells. Can you say a bit more about the kinetics of the response and whether it is different after lipopolysaccharide (LPS) or IL6?

Mantovani: IL6 is a weak inducer of PTX3, whereas it is the main inducer of the acute-phase proteins. In terms of its kinetics, this is one of the reasons we think it might be interesting in the clinic. It is clear that PTX3 levels increase much earlier than CRP. Take acute myocardial infarction, which is nice to study because it gives a start point for the inflammation. We did a cohort analysis involving measurements every 4 h. CRP will reach a peak between 35 and 48 h, whereas PTX3 will reach a peak in 6–8 h. We thought this was immediate early gene induction. But now that we have the neutrophil story we suspect that some of it comes from neutrophils. The kinetics are very different, as are the amounts.

McGreal: Have you looked at the alveolar macrophages and epithelial cells, with respect to PTX3 expression?

Mantovani: We haven't. We would have picked them up if it was constitutively present.

Romani: Dr Giovanni Salvatori (from Sigma-Tau in Rome) has done *in situ* histochemistry on epithelial cells from the bronchi. He found that PTX3 was locally present in the epithelial cells of the trachea (personal communication from Dr Salvatori).

McGreal: PTX3 has features which are analogous in many respects to the collectin family of C-type lectins; they are all multimeric, soluble, innate pathogen recognition molecules. Several of the collectins including MBL, SP-A and SP-D appear to play a role in the recognition of apoptotic cells as well as pathogens. Do you see a phenotype relating to apoptopic cells in the PTX3 knockout mice?

Mantovani: The data are as follows. *In vitro* it binds apoptotic cells late, after they become annexin positive and pick up PI. We looked at the interaction with dendritic cells and to our surprise we found that it inhibits uptake of apoptoptic cells. This is not the same as we see with conidia, for example, in which we see opsonization. We speculate that it may be involving molecules that are involved in editing self–non-self discrimination. The mice so far do not have an autoimmune phenotype.

Brown: My question is related to the expression data in the lung. Is it from the bronchoalveolar lavage (BAL) fluid in actual infection experiments?

Romani: No, it is intranasal.

Mantovani: Also the bacterial challenge is in the higher airway.

Speert: You made a fascinating observation that a single pattern recognition molecule can recognize so many different unrelated molecules. How does this happen? Is it through some conserved consensus sequence or a non-specific interaction?

Mantovani: We have the same question. There is obviously a division of labour between the two domains. We have started mapping ligands that interact with the N-terminal portion and the pentraxin domain. The question remains: how does the same molecule interact with a respectable affinity with so many different molecules? This is a general question applicable to pattern recognition. There was a review by David Stern talking about multiligands (Schmidt et al 2001). We need to have the structure and this would hopefully answer your question. At least, I am more at ease with the molecule now that we have a division of labour.

Speert: Are there loss of function mutations in humans that can explain, for example, why some people get much worse infections in cystic fibrosis? Have single nucleotide polymorphisms (SNPs) been identified?

Mantovani: There is one polymorphism in the molecule that we have reasons to suspect may be functional.

Speert: Recently, the observations in cystic fibrosis have been published. A modifier gene has been identified.

van Helden: I am also interested in the kinetics, but not the initial ones, rather the end-stage kinetics. How quickly does it drop? What happens if you have chronic inflammation? Does it remain constantly elevated?

Mantovani: The best kinetics data we have are from acute myocardial infarction. We first did a preliminary study and then a large series of studies. It goes down much faster than CRP. Intriguingly, we see it elevated as has been seen in tuberculosis, but these aren't spectacular elevations. I suspect that it can be elevated in chronic disease. Putting together all the published and unpublished data, the value of this molecule will be as an early marker. In dengue fever, for example, we have seen high levels. It comes up much earlier than CRP.

Segal: When you do binding studies, do you use the purified proteins?

Mantovani: The binding studies were all done with pure recombinant PTX3. This is done either with tagged PTX3 or untagged PTX3 with an antibody.

Segal: Is it more valid to do it in a tissue extract where there is competition among the different proteins? Then we could see which the predominant binding proteins are at the ambient concentrations.

Mantovani: We haven't done that experiment, although your point is a valid one. The interaction with C1q is ambivalent. In other words, if we have PTX3 bound on the substrate, then you get C1q and activation of the classic pathway. But if PTX3 is in the fluid phase, there is a paradoxical effect.

Latgé: Patients who are immunocompromised will be at risk of aspergillosis, so do you see a reduction in PTX3?

Mantovani: We have a small series of patients with aspergillosis. Some of them were high, others low for PTX3. In the next year we plan to do a prospective study in which we are asking the question in the clinic about whether PTX3 levels are predictive of susceptibility to aspergillosis.

Latgé: This would be even more important if you take bone marrow transplant patients to have a look at the PTX3 concentration, since this is one of the big risk factors in aspergillosis.

Mantovani: It is expressed in *Escherichia coli*. It is difficult to get an enterobacterium without OmpA for structural reasons, but we have done experiments that show differential binding to OmpA expressing and non-expressing bugs.

Lambrecht: I was puzzled by the fact that in ischaemic heart disease PTX3 levels predicted mortality. Another thing that happens in a high percentage of patients with acute myocardial infarction is that they get antibodies to their heart muscle. This is called Dressler syndrome and it can lead to myocarditis. You mentioned briefly that it can discriminate self from non-self. Do you think that in ischaemic tissues this might be the pattern recognition signal that leads to B cell responses as well? Is it an adjuvant when you add it to a harmless protein or to a self antigen?

Mantovani: We are testing the adjuvant function with one of the microbial molecules. We are doing the adjuvant experiment.

Brown: You showed an amplification with OmpA and PTX3. You alluded to the fact that it might be initiating complement activation and this would then target to the cell. There may be a pentraxin receptor. Is it complement-mediated directional amplification, or is it directly doing itself?

Mantovani: We don't know. We have evidence that at least in the air pouch system, which is a simple system, if you deplete complement you get an inhibition of the PTX3 amplification. This is a clean system where there is no reason to get opsonization, because OmpA is injected together with PTX3. In another model we have an indication that complement is involved, but we need to identify the receptors.

Schoub: Is the PTX3 elevation in dengue due to the dengue fever, or could it be used as a diagnostic marker for the onset of new lesions such as haemorrhagic fever? This would have important diagnostic implications in our situation for Congo Crimean Haemorrhagic fever. Having a diagnostic marker heralding new lesions would be extremely useful.

Mantovani: I am not a dengue fever expert, so this is difficult to answer. We got cases from Indonesia (Mairuhu et al 2005).

Finn: I don't know whether you have had a chance to collaborate with someone to look at samples from cancer patients. We have published that there is chronic inflammation in cancer. There are things that can be measured in the plasma. Is pentraxin something that could be involved here?

Mantovani: We have one cancer system, but we'd be interested in looking at more.

Finn: One of the reasons why we are interested in measuring levels of inflammation using various markers is because cancer growth usually results in the suppression of the adaptive immune system. Do your pentraxin knockout mice have otherwise normal adaptive immunity? Can you challenge them with antigens and get normal immune responses? Does it participate at proper levels in the induction of adaptive immunity?

Mantovani: We have no information in terms of conventional antigens and conventional immunization.

Finn: This goes back to the levels. The amount on its own can make a difference, aside from kinetics.

Romani: In PTX3-deficient mice there is a hyperinflammatory state, and they have a blunted adaptive immune response that can be reconstituted by adding PTX3 back.

Sheehan: I want to go back to your modelling of the interaction between haemagglutinin (HA) and pentraxin with TSG6. Does pentraxin bind hyaluronan itself, or does it do it through TSG6?

Mantovani: We made the observation that there is severe infertility in these mice. This is related to an abnormal assembly of the extracellular matrix of the cumulus oophorus which is essential for female fertility. It does not bind hyaluronic acid. It binds TSG6. It does not interfere with the recognition by TSG6 of hyaluronic acid by the link module, but it binds that area.

Sheehan: So hyaluronic acid and TSG6 can be found in lung secretions. I wondered whether there was a story there.

Mantovani: We have evidence of colocalization with TSG6. Our simplistic view is that there is this gigantic multimer which acts as a focal point for binding TSG6.

Williams: I was interested in what was going on in myocardial infarction. You showed PTX3 in the tissues. Is it bound to injured myocytes? Is it sparking off the recognition of the myocytes or did you only see it on other cells? What is it doing there in this type of sterile inflammation?

Mantovani: That is an interesting question, and it is one we have struggled with. We are relatively confident about the prognostic screening. We have done 740 patients. We now have a phenotype in the mouse. We think that in addition to being a marker, it has a modulatory function. It took a while to get a phenotype, but we have a phenotype in the myocardial infarction model in the mouse.

Steinman: You mentioned that the PTX3 knockout is sensitive to streptococcal pneumonia. Is this just an innate problem?

Mantovani: We are not sure about that. We are testing a possible adjuvant. I suspect it is innate.

Quesniaux: Do we know how these mice respond to tuberculosis?

Mantovani: No.

Speert: Can you administer this therapeutic once the infection is established?

Romani: Yes.

Quesniaux: What about pentraxins in *Drosophila*?

Mantovani: In *Drosophila* there is a phenotype called swiss cheese. They have never been challenged with microbes.

Lawn: Presumably, since long-chain pentraxins are released at sites of inflammation whereas short chain pentraxins such as C-reactive protein are released systemically, one might expect to find differential anatomical compartmentalization with higher concentrations of PTX3 at the actual sites of disease. Have you measured PTX3, for example in inflammatory pleural effusions?

Mantovani: We have found high levels in peritoneal effusions. I would assume you are right. What is seen in serum is just a minor reflection of what is happening in the tissues.

Romani: We have done the kinetics of PTX3 production in infected mice. It stays in the lung longer than in the serum.

References

Mairuhu AT, Peri G, Setiati TE et al 2005 Elevated plasma levels of the long pentraxin, pentraxin 3, in severe dengue virus infections. J Med Virol 76:547–552

Schmidt AM, Yan SD, Yan SF, Stern DM 2001 The multiligand receptor RAGE as a progression factor amplifying immune and inflammatory responses. J Clin Invest 108:949–955

How superoxide production by neutrophil leukocytes kills microbes

Anthony W. Segal

Centre for Molecular Medicine, University College London, 5 University Street, London WC1E 6JJ, UK

Abstract. Neutrophils represent the primary innate immune response to infection by bacteria and fungi which they ingest, kill and digest. Killing and digestion are dependent upon oxygen consumption by the NADPH oxidase which generates superoxide (O_2^-) in the phagocytic vacuole. Killing was thought to occur by free radical reactions of reactive oxygen species (ROS) with the microbes, or through the generation of HOCl by myeloperoxidase acting on H_2O_2. However, in knockout mice lacking the neutral proteases cathepsin G and elastase, these ROS do not kill microbes despite normal production of oxygen free radicals and halogenation. It turns out that the oxidase has another function. The passage of electrons is electrogenic and the charge generated across the wall of the phagocytic vacuole must be compensated if electron transport is to continue. This compensation is largely accomplished by the passage of Cl^-, which enters the vacuole from the granules, where it is present at a concentration of about 500 mM, into the cytosol. The pH of the vacuole is regulated by a Na^+/H^+ exchanger, NHE1, which pumps Na^+ out of the vacuole in exchange for cytosolic H^+ together with a flux of K^+ into the vacuole through the BKCa channel. These ion fluxes and pH changes serve to promote microbial killing and digestion by optimizing conditions for the action of the enzymes released from the cytoplasmic granules.

2006 Innate immunity to pulmonary infection. Wiley, Chichester (Novartis Foundation Symposium 279) p 92–100

Neutrophils are highly motile phagocytic cells that provide the first line of defence of the innate immune system against bacteria and fungi which they phagocytose and digest (Segal 2005). Phagocytosis is associated with a significant 'extra respiration of phagocytosis' which is non-mitochondrial and is essential for microbial killing. Killing and digestion are defective under anaerobic conditions and in chronic granulomatous disease (CGD); a profound immunodeficiency to bacterial and fungal infections was shown to be associated with failure of this respiratory burst.

The NADPH oxidase

The NADPH oxidase (recently reviewed Cross & Segal 2004, Vignais 2002) is composed of an electron transport chain in the membrane of the phagocytic vacuole that is activated by the co-coordinated activity of a number of cytosolic proteins.

Flavocytochrome b_{558} (NOX2) is the core electron-transporting component of the NADPH oxidase. It is distributed between the plasma membrane and membrane of the specific granules in neutrophils and is incorporated into the wall of the phagocytic vacuole where it forms a conduit for electrons to be pumped from NADPH in the cytosol onto oxygen to form O_2^- in the vacuole. It is a heterodimer composed of one molecule each of $gp91^{phox}$ and $p22^{phox}$.

$gp91^{phox}$ is composed of two major, and very different, domains. A hydrophilic C-terminal (282–570) portion contains the FAD and NADPH binding sites and the hydrophobic N-terminal half contains six membrane-spanning α-helices, amongst which are two haem prosthetic groups perpendicular to the plane of the membrane.

$p22^{phox}$ is a 194 amino acid (~21 kDa) protein with a hydrophobic, membrane-spanning N-terminus (1–132). It provides high-affinity binding sites for $p47^{phox}$ on a proline-rich domain (151–160) in the cytoplasmic hydrophilic C-terminus, and confers stability on $gp91^{phox}$.

Cytosolic phox proteins

$p67^{phox}$ (NOXA2), $p40^{phox}$ and $p47^{phox}$ (NOXO2) are rich in motifs involved in protein/protein interactions. These include SH3, TPR (tetratricopeptide repeat) PB1 and PX domains, and proline rich regions. Together with small Rho guanosine triphosphatase (GTPase) Rac 2, these proteins translocate to the wall of the phagocytic vacuole and interact with each other and the flavocytochrome, possibly changing its conformation and allowing access of the substrate, NADPH, to the active site close to FAD.

CGD and its molecular genetics

Chronic granulomatous disease (CGD) is a rare inherited disorder characterized by the absence of NADPH oxidase activity (Thrasher et al 1994). Phagocytes lacking NADPH oxidase activity are unable to kill bacteria and fungi efficiently, with the predicted consequence that these patients are profoundly immunodeficient, demonstrating frequent, severe, acute and chronic, often fatal, infections.

Defects in any one of four genes give rise to the known forms of CGD. *CYBB* (coding for $gp91^{phox}$, NOX2) is located on the X chromosome and accounts for about 65% of cases, exclusively in males (except in rare female carriers where there is extreme lyonization). The other three genes are all autosomal with *NCF1* ($p47^{phox}$ or NOXO2 protein) *NCF2* ($p67^{phox}$ or NOXA2) and *CYBA* ($p22^{phox}$) causing about 25%, 5% and 5% of cases respectively. No instances of CGD have been identified where a lesion of $p40^{phox}$ is causal.

A small subgroup of CGD patients have what is known as 'variant' CGD. In these cases there is partial loss of a protein or its function and often as much as 10% , and up to 30% (H. Malech, personal communication), of normal oxidase activity can be measured.

How neutrophils kill bacteria

Products of the oxidase and their implication in microbial killing

Initiation of NADPH oxidase activity coincides with degranulation, with a lag phase of about 20 seconds. It occurs after closure of the vacuole has occurred, and is limited to the plasma membrane comprising the vacuolar membrane.

Attention has focused upon the products of the oxidase themselves as the lethal agents.

O_2^- and H_2O_2

Neutrophils produce large amounts of O_2^-, estimated at about 1–4 M/l in the vacuole, although this value is probably too high and is more likely to be nearer 500 mM/l. The steady state concentration has been estimated to be in the micromolar range.

It is not clear what, if any, ROS other than O_2^- and H_2O_2, are produced in significant amounts in the vacuole.

Myeloperoxidase-mediated halogenation

Myeloperoxidase (MPO) constitutes about 5% of the total neutrophil protein and is thought to catalyse the H_2O_2 dependent oxidation of halides that can react with and kill microbes. MPO-mediated halogenation has been the accepted basis of microbial killing for several decades (Klebanoff 2005).

However, about 1 in 2000 of the general population are MPO deficient without any undue predisposition to infection, and the neutrophils of birds do not contain MPO.

Cytoplasmic granules and their contents

There are essentially two main granule types, the azurophils largely contain proteins and peptides directed towards microbial killing and digestion whereas the specific granules replenish membrane components and help to limit free radical reactions (Borregaard & Cowland 1997, Gullberg et al 1999). A number of reviews have recently covered this subject.

It was shown that phagocytosis was associated with discharge of the cytoplasmic granules into the vacuole. Alexander Fleming discovered and named lysozyme which he termed 'a remarkable bacteriolytic element found in tissues and secretions', including leukocytes. He showed that it lysed about two thirds of the bacteria he mixed with it.

Azurophil (or primary) granules contain MPO, cathepsin G, elastase and proteinase 3, all neutral proteases, defensins and about one third of the lysozyme. These contents are largely bound to an abundant matrix of sulfated mucopolysaccharide.

Specific (or secondary) granules contain lactoferrin, transcobalamin II, membrane components including flavocytochrome b558 of the NADPH oxidase approximately two thirds of the lysozyme and NGAL.

The influence of granule contents on killing of bacteria by ROS and hypochlorous acid

Staphylococcus aureus and *Escherichia coli* were incubated with 100 mM O_2^- or H_2O_2 or with 5 mM HOCl. O_2^- had little activity although H_2O_2 was more effective. HOCl was highly microbicidal. These killing effects were almost completely abrogated by the addition of granule contents (25 mg/ml) at a concentration of about 1/20th that in the vacuole. In addition, MPO (5 mg/ml) in saline completely blocked the killing activity of 100 mM H_2O_2 rather than enhancing it as would be expected if it were promoting killing through the generation of HOCl (Reeves et al 2003).

Conditions in the phagocytic vacuole

It is essential to have a clear understanding of the conditions in the phagocytic vacuole when attempting to define the killing mechanisms.

A heavily opsonized particle is taken up into the phagocytic vacuole within 20 seconds and killing is almost immediate (Segal et al 1981), the NADPH oxidase elevates the pH to about 7.8–8 in the first 3 minutes after phagocytosis after which it gradually falls to about 7.0 after 10–15 min (Segal et al 1981).

The human neutrophil has about 1000 granules, the contents of approximately 20 of which will be released into each vacuole, assuming that each neutrophil can engulf about 50 bacteria. These bacterial contents are squeezed onto the surface of the organism in very high concentrations (Reeves et al 2002, Segal et al 1980). It has been estimated that the granule protein makes up about 40% of the vacuolar volume, achieving protein concentrations of about 500 mg/ml.

Neutral proteases are essential for bacterial and fungal killing

Although the proposal that ROS would be toxic to living cells is intuitively attractive, it was never adequately tested under the conditions pertaining in the phagocytic vacuole. The opportunity was provided by the development of gene targeting which allowed a mouse model to be constructed which was lacking the major neutrophil proteases, elastase (NE), cathepsin G or both.

The loss of both NE and cathepsin G conferred as profound a defect of bacterial killing as was observed with the CGD mouse model (Reeves et al 2002).

The important point of these studies was that microbial killing was abolished despite a completely normal respiratory burst and normal levels of iodination, indicating that ROS and metabolites of the action of myeloperoxidase on H_2O_2 are not sufficient to kill these bacteria and fungi.

Thus it was clear that the combination of NADPH oxidase activity and neutral protease enzymes are require for microbial killing to take place, raising the question as to the connection between these two processes.

Activity of the NADPH oxidase alters the appearance of the contents of the phagocytic vacuole

The activity of the NADPH oxidase was found to alter the appearance of the contents of phagocytic vacuoles in electron micrographs of neutrophils examined soon after they had phagocytosed bacteria. These obvious structural differences coupled with the massive amounts of O_2^- injected into the vacuole, together with the fact that 10% of this amount of O_2^- was insufficient as shown by variant CGD, suggested that the oxidase was exerting some physicochemical influence on the granule contents rather than simply producing ROS or substrate for MPO. We therefore turned our attention to electron transport across the membrane and its consequences in terms of the movement of other ions.

Charge compensation across the vacuolar wall

The oxidase is electrogenic, and it transfers electrons across the vacuolar wall onto O_2 to form O_2^-, leaving protons in the cytoplasm. This charge separation depolarizes the membrane, which stops oxidase activity if the charge is not compensated.

The vacuole becomes alkaline despite the entry of acidic granule contents, indicating that the O_2^- and O_2^{2-} are consuming protons in the vacuole.

Charge compensation through the 'proton channel'

It had been believed that charge compensation was accomplished by protons entering the vacuole through either proton channels (DeCoursey 2003, Henderson &

Meech 2002) or through gp91^{phox} itself. Evidence for these channels was that protons channels are known to be blocked by Zn^{2+} and Cd^{2+} and it was thought that this also applied to the oxidase, and a complex explanation was offered for the fact that blockage of the oxidase required concentrations of Zn^{2+} three orders of magnitude higher.

We showed that the Zn^{2+} and Cd^{2+} were not inhibiting the oxidase but rather interfering with the assay of O^{2-} (Ahluwalia et al 2004).

Charge compensation through Cl^- channels

We have recently shown that the charge across the vacuolar membrane is compensated by Cl^- which enters the vacuole from the granules where it is present at a concentration of about 500 mM (unpublished).

Regulation of vacuolar pH

The attractive feature of charge compensation by protons was that it appeared to solve the problems of the changes in vacuolar pH induced by the oxidase. The passage of large amounts of O_2^- into the vacuole consumes all the free protons driving the pH up. At the same time protons have been released into the cytoplasm, which is acidified. The neutral proteases cathepsin G and elastase require a pH of between 7.0 and 9.0 for optimal activity and a mechanism is required for regulating it.

We have found that the primary regulation occurs through NHE1 (Zachos et al 2005) which exchanges Na^+ entering the vacuole from the granules, for cytosolic protons, and functions optimally at high vacuolar pH which it neutralizes.

K^+ enters the phagocytic vacuole through the large conductance Ca^{2+}-activated K^+ (BKCa) channel

K^+ enters the phagocytic vacuole through the large conductance Ca^{2+}-activated K^+ channel. This elevates the pH in the vacuole (Ahluwalia et al 2004).

The role of the NADPH oxidase induced ion fluxes is to activate the enzymes in the vacuole

Cathepsin G and elastase require a pH of between 7.0 and 9.0 to function efficiently and the oxidase elevates the pH from 6.0 to about 7.8 to 8.0.

A high concentration of K^+ solubilizes the cationic granule proteins by displacing them from the negatively charged sulfated proteoglycan granule matrix.

The granules contain 500 mM Cl^- which inhibits lysozyme, cathepsin G and elastase. These enzymes are activated when the Cl^- is pumped out of the vacuole to compensate the charge induced by electron transport into the vacuole (unpublished).

Acknowledgements

I thank the Wellcome Trust and CGD Research Trust for support.

References

Ahluwalia J, Tinker A, Clapp LH et al 2004 The large-conductance Ca2+-activated K+ channel is essential for innate immunity. Nature 427:853–858

Borregaard N, Cowland JB 1997 Granules of the human neutrophilic polymorphonuclear leukocyte. Blood 89:3503–3521

Cross AR, Segal AW 2004 The NADPH oxidase of professional phagocytes-prototype of the NOX electron transport chain systems. Biochim Biophys Acta 1657:1–22

DeCoursey TE 2003 Voltage-gated proton channels and other proton transfer pathways. Physiol Rev 83:475–579

Gullberg U, Bengtsson N, Bulow E, Garwicz D, Lindmark A, Olsson I 1999 Processing and targeting of granule proteins in human neutrophils. J Immunol Methods 232:201–210

Henderson LM, Meech RW 2002 Proton conduction through gp91phox. J Gen Physiol 120:759–765

Klebanoff SJ 2005 Myeloperoxidase: friend and foe. J Leukoc Biol 77:598–625

Reeves EP, Lu H, Jacobs HL et al 2002 Killing activity of neutrophils is mediated through activation of proteases by K+ flux. Nature 416:291–297

Reeves EP, Nagl M, Godovac-Zimmermann J, Segal AW 2003 Reassessment of the microbicidal activity of reactive oxygen species and hypochlorous acid with reference to the phagocytic vacuole of the neutrophil granulocyte. J Med Microbiol 52:643–651

Segal AW 2005 How neutrophils kill microbes. Annu Rev Immunol 23:197–223

Segal AW, Dorling J, Coade S 1980 Kinetics of fusion of the cytoplasmic granules with phagocytic vacuoles in human polymorphonuclear leukocytes. Biochemical and morphological studies. J Cell Biol 85:42–59

Segal AW, Geisow M, Garcia R, Harper A, Miller R 1981 The respiratory burst of phagocytic cells is associated with a rise in vacuolar pH. Nature 290:406–409

Thrasher AJ, Keep NH, Wientjes F, Segal AW 1994 Chronic granulomatous disease. Biochim Biophys Acta 1227:1–24

Vignais PV 2002 The superoxide-generating NADPH oxidase: structural aspects and activation mechanism. Cell Mol Life Sci 59:1428–1459

Zachos NC, Tse M, Donowitz M 2005 Molecular physiology of intestinal Na+/H+ exchange. Annu Rev Physiol 67. 411–443

DISCUSSION

Brown: What is the function of myeloperoxidase (MPO)? Does it have a role in bacterial killing?

Segal: If you do an experiment using radioactive iodide or chloride, allow neutrophils to phagocytose bacteria, and then run a 2D gel and see what is hot, one thing that is not hot is the bacterial proteins and structures. It is the autologous proteins. So you are not targeting the bacteria. MPO functions as a catalase, and I think its main function is to get rid of these free radicals. The chicken neutrophils I showed don't have any MPO at all, and about 1 in 1000 of the general population is MPO deficient and they don't show any immunodeficiency.

Brown: You said that 1 in 1000 lack this. Surely if it has a role in clearance of these free radicals there should be some defect in these individuals.

Segal: I think that these radicals have a damaging effect on the enzymes. If you don't get rid of them you will degrade these enzymes. So there is probably a marginal deficiency.

McGreal: What happens with this process in patients who are ventilated with a higher concentration of oxygen?

Segal: The oxidase is unaltered by high concentrations of oxygen. It will also function at about 1–2% of the normal ambient oxygen concentration.

Latgé: In the granules are there proteases that are active at low pH?

Segal: There are a few lysosomal-type enzymes. But these are present in tiny amounts, and they aren't sufficient to kill. One of the most important roles of the neutrophil is to digest. This is one of the problems in conditions such as CGD where the cells are not digesting the material; this is why you get granuloma formation and chronic inflammation. Part of the digestion process is by these acid hydrolases, which digest bits and pieces once most of the digestion has taken place by the neutral proteases. They don't seem to have a major role in the killing process.

Ryffel: pH changes induced enzyme activation. How is this pH change sensed? There is the notion of pH sensor proteins, which are G coupled proteins on the membrane. (Ludwig & Seuwen 2002, Ludwig et al 2003).

Segal: I think what is going on here is that electrons are being pumped across and then the charge is compensated. The protons are left outside the vacuole and the pH inside then has the potential to become very high. The granules release a lot of protons into the vacuole when they degranulate. In addition, a rise in pH switches on the Na^+/H^+ exchanger and there is this K^+ influx. The two are regulated differentially: as the pH rises the Na^+/H^+ exchanger is switched on, and if it goes too far the K^+ is exchanged instead of the Cl^-.

E Sim: What is the phenotype of the knockout mice you mentioned?

Segal: They are fine.

E Sim: Have there been any human examples of modifications in cathepsins or elastins?

Segal: We haven't looked for phenotypes. They should be present. The ordinary CGD mouse also has a normal phenotype unless you challenge it.

E Sim: Many of my colleagues think that Ca^{2+} is everything. You haven't mentioned Ca^{2+} at all. Are there any ryanodine receptors involved?

Segal: The answer is that I don't know. Ca^{2+} is involved in the K^+ channel; the BK is a Ca^{2+}-dependent channel. It is complicated. What happens is that when you have your vacuole you pump so many electrons across, you leave a little layer of protons just around the vacuole. This seems to produce a zone of elevated Ca^{2+}.

Finn: I know that neutrophils are your favourite cells, but natural killer T (NKT) cells use granular exocytosis as a means of killing other cells. These enzymes also probably need activation. Is your process applicable to the granules made by T cells and NK cells?

Segal: I'm sure people haven't thought enough about the activation mechanisms. I would like to think that they are, but I have enough problems at the moment with just the neutrophils.

Brown: What about macrophages?

Segal: They start off as monocytes with an oxidase. You put them into culture and all the oxidase components are lost. Then you activate them, but no one seems to have looked at what this activation represents in terms of protein granule synthesis and oxidase component synthesis.

Romani: What sort of neutrophils do you study?

Segal: They are human.

Romani: Have you seen differences among neutrophils from different sources?

Segal: They all have their respiratory burst. This was initially described in dogs in 1938. We have looked at mouse neutrophils. They all seem to be the same.

Quesniaux: How stable is the level of MPO over the lifespan of the neutrophil?

Segal: We only see the end cells. They come out after they have been in the bone marrow for about six days, stay in the circulation for six hours and then enter the tissue for 1–2 days. We only see them for the six hours. MPO is about 5% of the total protein of the cell.

Gordon: It is interesting that during monocyte differentiation MPO tends to be lost and isn't replaced. Many mature macrophages have lost it.

References

Ludwig MG, Seuwen K 2002 Characterization of the human adenylyl cyclase gene family: cDNA, gene structure, and tissue distribution of the nine isoforms. J Recept Signal Transduct Res 22:79–110

Ludwig MG, Vanek M, Guerini D et al 2003 Proton-sensing G-protein-coupled receptors. Nature 425:93–98

Linking innate to adaptive immunity through dendritic cells

Ralph M. Steinman

The Rockefeller University, New York, NY 10021, USA

Abstract. The function of dendritic cells (DCs) in linking innate to adaptive immunity is often summarized with two terms. DCs are sentinels, able to capture, process and present antigens and to migrate to lymphoid tissues to select rare, antigen-reactive T cell clones. DCs are also sensors, responding to a spectrum of environmental cues by extensive differentiation or maturation. The type of DC and the type of maturation induced by different stimuli influences the immunological outcome, such as the differentiation of Th1 vs. Th2 T cells. Here we summarize the contributions of DCs to innate defences, particularly the production of immune enhancing cytokines and the activation of innate lymphocytes. Then we outline three innate features of DCs that influence peripheral tolerance and lead to adaptive immunity: a specialized endocytic system for antigen capture and processing, location and movements *in vivo*, and maturation in response to an array of stimuli. A new approach to the analysis of DC biology is to target antigens selectively to maturing DCs *in vivo*. This leads to stronger, more prolonged and broader (many immunogenic peptides) immunity by both T cells and B cells.

2006 Innate immunity to pulmonary infection. Wiley, Chichester (Novartis Foundation Symposium 279) p 101–113

Innate defenses provided by dendritic cells

There are several ramifications to the term 'innate immunity' for dendritic cells (DCs). The traditional consequence, as discovered by Metchnikoff with macrophages, is innate 'defence', that is processes that provide rapid resistance to infection. Metchnikoff discovered that phagocytosis could provide resistance to infection. DCs also take up particles, but in relatively small numbers, and the cells are proficient at antigen processing and presentation rather than microbial killing. However, there is a new entity in inflammation termed 'tip DC', which is a cell identified by Pamer and colleagues during murine Listeriosis (Serbina et al 2003) and more recently in lesional skin of patients with psoriasis. The hallmark of this cell is the abundance of tumour necrosis factor (TNF) and inducible nitric oxide synthase (iNOS) (hence the term TNF and iNOS producing DC), and these proteins contribute to host defence.

There are other forms of innate defences, many of which are energized through Toll-like receptors (TLRs) or intracellular microbial sensors like RIG-I and NOD proteins. Cultured macrophages and DCs share the capacity to make large amounts of cytokines upon microbial challenge, particularly ligation of TLRs. *In vivo* studies in mice have shown that DCs can be the major source for two of these, interleukin (IL)12 and type I interferons (Dalod et al 2002, Reis e Sousa et al 1997), which provide innate defences and also act as adjuvants for adaptive immunity.

The production of inflammatory cytokines by DCs is induced by another pathway for innate defence, innate lymphocytes. DCs seem to be the main cells that activate and expand innate lymphocytes, i.e. natural killer (NK), natural killer T (NKT) and γδ T cells (Munz et al 2005). For DCs and NK cells, the two cells are now known to be juxtaposed in the T cell areas of lymphoid tissues (Ferlazzo et al 2004). One consequence of the interaction of DCs with innate lymphocytes is the production of TNF and IL12, but interestingly, this induction is independent of the MyD88 adapter protein for TLRs (Fujii et al 2003).

A second meaning of the term innate is 'built in'. DCs have such properties that are rapidly available but often do not directly contain an infection. Instead, these features allow DCs to link innate with adaptive immunity, and will now be outlined.

Summary. The field of innate immunity has mushroomed beyond Metchnikoff's phagocytosis to include numerous anti-microbial pathways. These include complement, a host of anti-microbial opsonins and peptides, cytokines, chemokines and innate lymphocytes. DCs have important roles in innate immunity, particularly with regard to cytokine production and mobilization of innate lymphocytes, and DCs have specialized innate features that lead to tolerance and adaptive immunity.

The innate features of DCs that control adaptive immunity—position and homing

DCs are positioned at mucosal surfaces like the airway and are able to home to the T cell areas of draining mucosal associated lymphoid organs, e.g. the lymph nodes in the chest or mediastinum. The distribution and movement of DCs allows the cells to sample environmental and self proteins in the steady state for purposes of tolerance, and under conditions of perturbation, to present microbial and other antigens for purposes of allergy and immunity (Brimnes et al 2003).

An active field of current research involves the types of DCs in the lung and their distribution there. The field was energized by Holt and colleagues, particularly in respiratory epithelium, where DCs are found within and just below the epithelium (Holt et al 1987). There is an important contrast between DC function and the traditional barrier function of epithelia. In the steady state, in the ostensible absence of infection and inflammation, peripheral DCs continuously capture environmental proteins, e.g. from the airway and intestinal lumen (Brimnes et al 2003, Liu & MacPherson 1993). DCs can produce the components of the occluding epithelial junction, e.g. claudens and occludens, and this could allow dendrites to penetrate between epithelial cells without altering the epithelial barrier (Rescigno et al 2001).

The identification of DCs in sections of lymphoid tissues has been achieved by a combination of criteria in addition to large irregular cell shapes. These include high expression of MHC class II and CD11c integrin, a lack of lymphocyte and macrophage markers, and expression of receptors for antigen uptake (below). The distribution of DCs has now been visualized by two-photon microscopy of living tissue. Migrating DCs arrive in the T cell area where they efficiently select T cells specific for presented antigens (Mempel et al 2004). The DCs join a network that is present in the steady state, as visualized using a 'green DC' mouse in which a fluorescent protein is driven by a CD11c promoter that is most active in DCs (Lindquist et al 2004). Stable contacts develop when antigen bearing DCs encounter their cognate T cells, and these contacts persist for at least 18 hours. These contacts are apparent in the steady state, when DCs are tolerogenic, and upon maturation when immunity develops.

Summary. A significant innate property of DCs, which facilitates the initiation of adaptive immunity, is their ready access to antigens at mucosal surfaces and their capacity to move to lymphoid tissues. There, DCs constitute just a few percent of total cells, but their size and pervasive cell shape puts them in a position to scan T cells circulating through lymphoid tissues and to select antigen-specific clones.

The innate features of DCs that control immunity—the endocytic system

DCs express an array of receptors that mediate endocytosis. Many of these are also expressed by macrophages and include an array of calcium-dependent lectins. Macrophages are proficient at endocytosis over long periods and catabolism to amino acids, whereas DCs take up antigens over short periods of time and are designed to efficiently process these to peptides for binding to MHC products. In most instances, relatively little is known about self and foreign ligands for these receptors. DC-SIGN/CD209, which is expressed on monocyte derived DCs, is a well-studied exception. It recognizes mannose and fucosyl residues on several pathogens including HIV, CMV, Ebola and dengue viruses, yeasts, and certain *Leishmania* (Figdor et al 2002).

The uptake functions of DC receptors have been studied in three ways. A role in uptake is predicted by the presence of characteristic coated pit localization sequences in the cytosolic domains, as well as motifs for targeting within the cell (Mahnke et al 2000). In some cases, ligands are followed into the cell, while in others, antibodies are used as surrogate ligands. Antibodies are providing a new way to study endocytosis receptors within intact lymphoid organs. Anti-receptor antibodies serve as delivery vehicles for antigens that are either chemically coupled or genetically introduced into the antibody. Antigen fusion antibodies to DEC-205/CD205 efficiently induce immunity in naïve mice when given together with a stimulus for DC maturation (Bonifaz et al 2004, Boscardin et al 2006, Trumpfheller et al 2006).

At the mRNA level, DCs express several receptors implicated in the uptake of dying cells. For irradiated tumour cells, dying autologous splenocytes and allogeneic cells killed by NK cells, a perplexing feature is that uptake is restricted to CD8α+ DCs, even though both CD8α− and CD8α+ subsets express potential uptake receptors (Iyoda et al 2002). The capacity of DCs to take up dying cells is a pivotal research area because so many clinically relevant antigens could gain access to DCs in this way, e.g. autoantigens in tissues undergoing turnover, tumours, transplants and infections.

DCs express FcγRs (and FcεRs) and thereby present immune complexes and antibody coated tumour cells on both MHC class I and II. Importantly, FcγRs influence the state of DC maturation. FcγRs associated with activating ITAM sequences stimulate, while FcγRs with inhibitory ITIM sequences block maturation (Dhodapkar et al 2005, Kalergis & Ravetch 2002). FcRs also represent a pivotal area for future research because of their number, ability to discriminate subtypes of immunoglobulin and immune complexes, and significant potential to influence DC function.

There is considerable potential to receptor function beyond the classical role in uptake and processing onto MHC class II. Uptake receptors can associate with other signalling molecules, like TLR2 and Syk for Dectin-1. Second, individual receptors can follow distinct trafficking paths dictated by cytosolic domain sequences. DEC-205/CD205 has a stretch of three acidic amino acids that allows this receptor (uniquely at this time) to target and slowly recycle through MHC class II^+ late endosomes (Mahnke et al 2000). An enigmatic consequence of antigen uptake is cross presentation on MHC class I, which is evident for captured dying cells, immune complexes, and DEC-205 ligands. It is not clear how this cross presentation comes about. Individual receptors also can be expressed on distinct subsets of DCs. Altogether, DCs are endowed with numerous receptors, providing a means to take up many ligands and carry out distinct 'post uptake' outcomes.

The endocytic system, not just the repertoire of uptake receptors, is proving to be a distinctive innate feature of DCs. *Ex vivo* studies reveal a significant, even unique regulation at several levels (Trombetta & Mellman 2005). To begin, uptake by pinocytosis and phagocytosis can be curtailed when DCs mature, through inactivation of a required rhoGTPase. This should limit presentation to antigens captured in the periphery, when DCs are immature and responding to innate stimuli, and not self antigens taken up following arrival in the T cell areas. The lysosomes of certain immature DCs are unusual relative to other cells, in that proteins are degraded slowly. This reflects two features: a relatively high intravacuolar pH and a lack of proteases (Delamarre et al 2005). When DCs receive a maturation stimulus, a proton pump assembles on the vacuolar membrane, the pH falls to 4.5–5.0, and proteolysis begins. Following their formation within maturing DCs, peptide–MHC complexes move within distinct non-lysosomal vesicles to the surface (Turley et al 2000). These transport organelles also contain the costimulatory molecule CD86, which then remains clustered with peptide–MHC at the DC surface (Turley et al 2000). Clustering of T-cell receptor (TCR) ligands and costimulators possibly accounts for efficient and prolonged stimulation of the TCR and CD28 on T cells.

The CD1 family of non-classical MHC class I molecules recognizes glycolipids. DCs are a major site for the expression of CD1a (Langerhans cells), CD1b and c (dermal dendritic cells, other interstititial DCs and myeloid DCs) and CD1d (most DCs). CD1d presents glycolipids to the invariant TCR on NKT lymphocytes. The glycolipids are derived from endogenous, microbial, allergic and synthetic sources. An important feature of CD1d presentation is that it leads to changes in DC function. For example, a single dose of the synthetic glycolipid, α-galactosyl ceramide (αGal-Cer), leads to DC maturation and to Th1 $CD4^+$ and $CD8^+$ T cell responses to protein antigens (Fujii et al 2003). On the other hand, multiple doses of the glycolipid dampen immunity, which can involve the formation of regulatory IL10 producing DCs (Kojo et al 2005). Since NKT cells, like T cells can differentiate along functionally distinct pathways, the capacity of DCs to present glycolipids provides another dimension to control immunity.

Summary. A distinct innate feature of DCs is their endocytic system, which helps to explain their efficient translation of innate to adaptive immunity. DCs express many uptake receptors. Some already are known to lead to presentation on both MHC class I and II products. Relatively low doses of antigen often suffice for DCs to present antigens to T cells, suggesting that adsorptive uptake is taking place and/or that processing is efficient following uptake. The endocytic system of DCs is peculiar in its regulation at many levels during maturation, including the expression of uptake receptors, formation of endocytic vacuoles, and the acidity and therefore activity of the vacuolar system. Overall, the DC endocytic system is specialized for antigen presentation rather than clearance and scavenging, as is the case for macrophages and granulocytes.

The innate features of DCs that control immunity—maturation

The term maturation was used to describe the differentiation of DCs required for the induction of immunity. This usage is still valuable to describe the extensive changes in response to TLR ligands and other maturation stimuli, e.g. CD40 ligation, necrotic cells, certain cytokines, innate lymphocytes and immune complexes. The initial experiments involved Langerhans cells (LCs), which were weak stimulators of the mixed leukocyte reaction and other T cell responses. The LCs only became strong stimulators after culture in GM-CSF. This maturation of T cell stimulatory function was accompanied by extensive differentiation i.e. the appearance and loss of DC markers and development of a highly 'dendritic' morphology. Strikingly, freshly isolated LCs could capture antigens for presentation to activated T cells, but the mature LCs did not (Romani et al 1989). These observations first distinguished two broad requirements for immunity: an antigen capture and processing step carried out by immature LCs, and an accessory (later 'costimulatory') function carried out by DCs that were surprisingly incapable of antigen capture.

Shortly thereafter, when antibodies to CD86 became available, it was recognized that maturing DCs up-regulated expression of B7-2/CD86 more rapidly and to higher levels than LPS-stimulated macrophages and B cells (Inaba et al 1994). Many scientists then considered heightened B7 expression to be synonymous with maturation, but this is not the case. B7 expression is noted on DCs that are immature in many respects, even DCs that induce tolerance. During inflammation, DCs up-regulate B7 family members to high levels but the DCs can remain functionally immature, i.e. nonimmunogenic, unless additional cues like CD40 ligation are received (Fujii et al 2004). Likewise CD86 is expressed by 'bystander' DCs, which are DCs with appropriate MHC products but lacking the MyD88 adaptor to respond to TLR ligands. These bystander DCs were hypoactive in inducing immunity (Sporri & Reis e Sousa 2005). Thus *heightened* B7 expression usefully monitors DC responses to inflammation or infection, e.g. to cytokines like TNFα, but it is not *equivalent* to maturation (Fujii et al 2004).

Maturation typically entails, in addition to higher expression of B7s and CD40, the production of chemokines and cytokines and the expression of other costimulators, e.g. CD54, CD58, TNF family members, notch ligands and T-bet transcription factor. These changes represent cues that DCs are differentiating, but more research is needed to determine how each aspect of DC maturation contributes to the induction T cell differentiation and memory.

Importantly, distinct stimuli allow DCs to initiate distinct responses. A good example would be the myeloid DCs in human blood. When these cells encounter thymic stromal lymphopoietin (TSLP) or CD40L, the cells look like similar mature DCs. They have heightened MHC class II and B7 expression and are highly dendritic. However, TSLP DCs cause naïve T cells to differentiate into inflammatory Th2 cells that produce TNF in addition to IL4, 5 and 13, while CD40L DCs cause naïve T cells to differentiate into Th1 cells (Soumelis et al 2002). Deeper analysis reveals that TSLP DCs make distinct chemokines from CD40L DCs and fail to make inflammatory cytokines like IL1, IL6 and IL12. One of the major enigmas in DC biology is to characterize the responses to microbial products, like schistosome egg antigen, or certain allergens (Traidl-Hoffmann et al 2005), which allow DCs to induce more classical 'noninflammatory' Th2 T cells. In other words, distinct maturation stimuli can influence the outcome of DC–T cell interactions.

DCs respond rapidly to Toll-like receptor ligands (TLRs), germline-encoded receptors for microbial products. Two cytokines produced as a result of TLR signalling have significant immune enhancing effects. IL12, whose production is enhanced by the transcription factor IRF5 (Takaoka et al 2005), acts on $CD4^+$ T cells to enhance Th1 differentiation. Type I interferons, whose production is enhanced by the transcription factor IRF7 (Honda et al 2005), act on $CD8^+$ T cells (Kolumam et al 2005) and B cells (Le Bon et al 2001) to enhance CTL, antibody formation and memory. The differentiation of helper T cells can be influenced by the type of TLR ligand. CpG DNA, a TLR9 ligand, and imiquimod, a TLR7 ligand, enhance Th1 type immune responses. The TLR2 ligand, Pam3Cys, and the TLR5 ligand, flagellin, in contrast can induce Th2 type responses *in vitro*. The effects of microbial ligands may depend on the stage of DC development, e.g. lipopolysaccharide and bacteria can *inhibit* the differentiation of DCs from monocytes *in vivo* (Rotta et al 2003), but *enhance* differentiation of immature DCs. The field is now poised to consider DC responses to TLR ligands *in vivo* and an in-depth analysis of consequences with respect to immunity, memory and tolerance.

Summary. DCs are sensors, responding to an array of environmental stimuli that range from microbial ligands for TLRs, CD40 ligation, necrosis, innate lymphocytes and immune complexes. The response entails extensive typically terminal differentiation, called maturation. The maturation program varies with the stimulus, and the consequences for lymphocytes are likewise different. Maturation allows DCs to influence T cell differentiation along different types of Th1 and Th2 pathways, and it is proving to be important in the genesis of different types of peripheral tolerance and memory. Maturation comprises many components that lead to adaptive immunity: prolonged presentation of lymphocyte ligands such as MHC–peptide complexes, numerous membrane costimulators, and production of large amounts of enhancing cytokines like IL12 and type I interferons. The latter provide innate defence and act on both DCs and lymphocytes to energize adaptive immunity. At this time, much of the literature involves *ex vivo* studies, TCR transgenic T cells, model antigens and limited readouts of T cell function. A new approach to DC function *in vivo*, and to harness this knowledge in vaccine biology and other challenging clinical areas, is to target antigens selectively to maturing DCs. In naïve animals, this leads to stronger, more prolonged and broader (many immunogenic peptides) immune responses, by both T cell (Bonifaz et al 2004, Trumpfheller et al 2006) and B cells (Boscardin et al 2006).

References

Bonifaz LC, Bonnyay DP, Charalambous A et al 2004 In vivo targeting of antigens to the DEC-205 receptor on maturing dendritic cells improves T cell vaccination. J Exp Med 199:815–824

Boscardin SB, Hafalla JC, Masilamani RF et al 2006 Antigen targeting to dendritic cells elicits long-lived T cell help for antibody responses. J Exp Med 203:599–606

Kamphorst AO, Zebroski HA, Rai U et al 2003 Influenza virus-induced dendritic cell maturation is associated with the induction of strong T cell immunity to a coadministered, normally nonimmunogenic protein. J Exp Med 198:133–144

Dalod M, Salazar-Mather TP, Malmgaard L et al 2002 Interferon α/β and interleukin 12 responses to viral infections: pathways regulating dendritic cell cytokine expression in vivo. J Exp Med 195:517–528

Delamarre L, Pack M, Chang H, Mellman I, Trombetta ES 2005 Differential lysosomal proteolysis in antigen-presenting cells determines antigen fate. Science 307:1630–1634

Dhodapkar KM, Kaufman JL, Ehlers M et al 2005 Selective blockade of inhibitory Fcγ receptor enables human dendritic cell maturation with IL-12p70 production and immunity to antibody-coated tumor cells. Proc Natl Acad Sci USA 102:2910–2915

Ferlazzo G, Thomas D, Lin SL et al 2004 The abundant NK cells in human secondary lymphoid tissues require activation to express killer cell Ig-like receptors and become cytolytic. J Immunol 172:1455–1462

Figdor CG, van Kooyk Y, Adema GJ 2002 C-type lectin receptors on dendritic cells and Langerhans cells. Nat Rev Immunol 2:77–84

Fujii S, Shimizu K, Smith C, Bonifaz L, Steinman RM 2003 Activation of natural killer T cells by α-galactosylceramide rapidly induces the full maturation of dendritic cells in vivo and thereby acts as an adjuvant for combined CD4 and CD8 T cell immunity to a co-administered protein. J Exp Med 198:267–279

Fujii S, Liu K, Smith C, Bonito AJ, Steinman RM 2004 The linkage of innate to adaptive immunity via maturing dendritic cells in vivo requires CD40 ligation in addition to antigen presentation and CD80/86 costimulation. J Exp Med 199:1607–1618

Holt PG, Schon-Hegrad MA, Oliver J 1987 MHC class II antigen-bearing dendritic cells in pulmonary tissues of the rat. Regulation of antigen presentation activity by endogenous macrophage populations. J Exp Med 167:262–274

Honda K, Yanai H, Negishi H et al 2005 IRF-7 is the master regulator of type-I interferon-dependent immune responses. Nature 434:772–777

Inaba K, Witmer-Pack M, Inaba M et al 1994 The tissue distribution of the B7–2 costimulator in mice: abundant expression on dendritic cells in situ and during maturation in vitro. J Exp Med 180:1849–1860

Iyoda T, Shimoyama S, Liu K et al 2002 The $CD8^+$ dendritic cell subset selectively endocytoses dying cells in culture and in vivo. J Exp Med 195:1289–1302

Kalergis AM, Ravetch JV 2002 Inducing tumor immunity through the selective engagement of activating Fcγ receptors on dendritic cells. J Exp Med 195:1653–1659

Kojo S, Seino K, Harada M et al 2005 Induction of regulatory properties in dendritic cells by Vα14 NKT cells. J Immunol 175:3648–3655

Kolumam GA, Thomas S, Thompson LJ, Sprent J, Murali-Krishna K 2005 Type I interferons act directly on CD8 T cells to allow clonal expansion and memory formation in response to viral infection. J Exp Med 202:637–650

Le Bon A, Schiavoni G, D'Agostinio G et al 2001 Type I interferons potently enhance humoral immunity and can promote isotype switching by stimulating dendritic cells in vivo. Immunity 14:461–470

Lindquist RL, Shakhar G, Dudziak D et al 2004 Visualizing dendritic cell networks in vivo. Nat Immunol 5:1243–1250
Liu LM, MacPherson GG 1993 Antigen acquisition by dendritic cells: intestinal dendritic cells acquire antigen administered orally and can prime naive T cells in vivo. J Exp Med 177:1299–1307
Mahnke K, Guo M, Lee S et al 2000 The dendritic cell receptor for endocytosis, DEC-205, can recycle and enhance antigen presentation via major histocompatibility complex class II-positive lysosomal compartments. J Cell Biol 151:673–683
Mempel TR, Henrickson SE, Von Andrian UH 2004 T-cell priming by dendritic cells in lymph nodes occurs in three distinct phases. Nature 427:154–159
Munz C, Steinman RM, Fujii S 2005 Dendritic cell maturation by innate lymphocytes: coordinated stimulation of innate and adaptive immunity. J Exp Med 202:203–207
Reis e Sousa C, Hieny S, Scharton-Kersten T et al 1997 *In vivo* microbial stimulation induces rapid CD40L-independent production of IL-12 by dendritic cells and their redistribution to T cell areas. J Exp Med 186:1819–1829
Rescigno M, Urbano M, Valzasina B et al 2001 Dendritic cells express tight junction proteins and penetrate gut epithelial monolayers to sample bacteria. Nat Immunol 2:361–367
Romani N, Koide S, Crowley M et al 1989 Presentation of exogenous protein antigens by dendritic cells to T cell clones: intact protein is presented best by immature, epidermal Langerhans cells. J Exp Med 169:1169–1178
Rotta G, Edwards EW, Sangaletti S et al 2003 Lipopolysaccharide or whole bacteria block the conversion of inflammatory monocytes into dendritic cells in vivo. J Exp Med 198: 1253–1263
Serbina NV, Salazar-Mather TP, Biron CA, Kuziel WA, Pamer EG 2003 TNF/iNOS-producing dendritic cells mediate innate immune defense against bacterial infection. Immunity 19:59–70
Soumelis V, Reche PA, Kanzler H et al 2002 Human epithelial cells trigger dendritic cell-mediated allergic inflammation by producing TSLP. Nat Immunol 3:673–680
Sporri R, Reis e Sousa C 2005 Inflammatory mediators are insufficient for full dendritic cell activation and promote expansion of $CD4^+$ T cell populations lacking helper function. Nat Immunol 6:163–170
Takaoka A, Yanai H, Kondo S et al 2005 Integral role of IRF-5 in the gene induction programme activated by Toll-like receptors. Nature 434:243–249
Traidl-Hoffmann C, Mariani V, Hochrein H et al 2005 Pollen-associated phytoprostanes inhibit dendritic cell interleukin-12 production and augment T helper type 2 cell polarization. J Exp Med 201:627–636
Trombetta ES, Mellman I 2005 Cell biology of antigen processing in vitro and in vivo. Annu Rev Immunol 23:975–1028
Trumpfheller C, Finke JS, Lopez CB et al 2006 Intensified and protective $CD4^+$ T cell immunity in mice with anti-dendritic cell HIV gag fusion antibody vaccine. J Exp Med 203: 607–617
Turley SJ, Inaba K, Garrett WS et al 2000 Transport of peptide-MHC class II complexes in developing dendritic cells. Science 288:522–527

DISCUSSION

Hussell: Tolerance occurs in the absence of flu. Do you think this is because flu breaks down the epithelial barrier so you get presentation by dendritic cells (DCs) in the presence of flu, but presentation by alveolar macrophages in its absence?

Steinman: In the absence of flu the only presenting cell we see in the lymph node is the DC.

Hussell: Are you giving the ovalbumin intranasally?

Steinman: It is aerosolized. In the presence of flu the only cell we see presenting ovalbumin is in the CD11c positive fraction. This doesn't prove what is happening in the intact animal, but when we isolate the cells and try to characterize them, it seems to be mainly the DC that is presenting antigen.

Mantovani: Is DEC-205 silent in terms of giving a signal?

Steinman: We don't see signalling motifs in the cytosolic domain, and we don't see changes in the various surface markers that we follow, or in the production of cytokines or chemokines. So it looks silent.

Brown: Can you comment on how you have tried to identify the endogenous ligand of DEC-205?

Steinman: We made soluble DEC-205 and applied that to either sections or to libraries. The glycan libraries were assessed with Ten Feizi in London. We didn't see any specific binding.

Brown: Have you injected the soluble form of this receptor?

Steinman: No. If we add a lot to the section we see binding but it is all over the place.

Brown: Is it cleaved?

Steinman: I don't know.

Sheehan: Generally, individual oligosaccharide binding to protein is a weak process. It becomes much stronger when a collective assembly of oligosaccharides is involved. You have all these lectins in a line, and for all you know they could be targeted to a complex, higher-organized structure of oligosaccharides on a surface. Were your experiments done to recognize that fact, or were these individual oligosaccharide ligands, to which your protein was exposed?

Steinman: We used a number of glycoproteins and did not see binding.

Gordon: You should probably try some higher-affinity ligands for multimerizing the DEC-205.

Sheehan: The glycoproteins it is recognizing could be very large molecules with multiple carbohydrate ligands on them. You would have missed these entirely.

Brown: DEC-205 may also recognise a microbial ligand.

Steinman: If we used DEC knockout mice we could probe this by looking at uptake of specific organisms.

Lambrecht: In the original Ova model where you injected DEC-205-OVA and then showed that in the absence of maturation you were inducing a tolerogenic response, it would also be interesting to look at the vaccinia Gag system, where you are coming back with your antigen but this time in an immunogenic complex. If you gave DEC-205-Gag in the absence of poly IC and CD40 ligands, can you make viral infection worse if you didn't induce any DC maturation at the time of DEC-205-Gag injection?

Steinman: In the assays we use we are giving a high dose of vaccinia Gag. The animals who receive nothing lose weight quickly. I don't think we would have picked up tolerance with this type of challenge.

Lambrecht: This is highly relevant for tumour immunity. If you try to induce DC maturation *in vivo* in a host that carries a tumour that's immunosuppressive, what will happen to your DC maturation? Could targeting your antigen to immature DCs be detrimental?

Steinman: This is true of all vaccines. How do we know that DNA vaccines aren't tolerizing the CD4 cells?

Didierlaurent: Are you sure that the antigens target to a specific subset of DCs, and this is linked to a memory response for T cells? What is so special about these DCs?

Steinman: We don't know yet whether it is necessary to target this particular DEC-205 receptor or the $CD8^+$ subset that expresses it. We have started targeting the other $CD8^-$ subset. There is a monoclonal called 33D1 that lets us do this. The experiments are not out to the memory phase yet. The $CD8^-$ subset doesn't present antigen well to CD8 cells, so there is a big difference between subsets. It could be that it is the specific receptor in addition to the subset that makes a difference. The other thing that is special about the $CD8^+$ subset is that it is the one that makes more IL12, but both the systems that I described work fine in an IL12 knockout.

Romani: When you measure memory T cells, can you comment on the finding that you do not see any IL10 being produced?

Steinman: We have looked primarily for IL4 which is not detectable by Elispot or FACS, and then we looked for IL10 which is also not detectable. It is very different from a traditional immunization.

Quesniaux: Is DEC-205 expressed on other cells?

Steinman: It is abundant on the DCs in the T cell areas, but it is not specific. All leukocytes have a bit of DEC, as best we can tell. Many epithelia have a lot of DEC-205, primarily on the basal side. But when we target DEC-205 with the antibody, it mainly goes to the DCs in the T cell area.

Walzl: The maturation path of DCs is obviously important in the outcome of infections. Do you think that in the context of chronic infection with a lot of dying cells around, those DCs will be presenting *Mycobacterium tuberculosis* antigens, for example, and induce tolerance?

Steinman: I feel DCs are really good at presenting dying cells. When you work out how much antigen you are delivering and then the immune response you are getting, it is really efficient. But the outcome is what I can't predict because I don't know how the infection will influence maturation. The idea is that the dying cells we study don't deliver any information other than antigens, but this could be wrong.

Finn: The type of dying cell could be important. While necrosis versus apoptosis isn't as black and white a distinction as some think, there is some evidence that cells make certain molecules when dying by necrosis that stimulate maturation of DC and give an immunogenic stimulus. From Ruslan Medzhitov's work at Yale it is clear that inert antigens will be trafficked differently than pathogens, into different lysozomes, and the result will be different. It is not all in the type of the DC.

Ryffel: You used irradiation of the tumour cells in your model. But if you 'sensitize' the tumour cells differently, such as with chemotherapeutic agent doxorubicin as described recently (Casares et al 2005) do you have the same effect?

Steinman: We have started looking at other ways of killing the tumour cell. We are still not seeing any difference from irradiation with chemotherapeutics. We still don't see the tumour cell maturing the DC on its own; there is still a requirement for the maturation stimulus. So far we haven't been able to change what I have just shown with simple irradiation.

Ryffel: You showed this αGal-Cer effect in the tumour model. Did you see this stimulation in the vaccine protection? The poly IC anti-CD40 is used as adjuvant. Would this be an alternative?

Steinman: I suspect not, because the αGal-Cer works best through the i.v. route. If it is given subcutaneously, this doesn't affect DC function. Since we want our vaccines to be given subcutaneously, I suspect αGal-Ser will not be a good adjuvant. In terms of which adjuvant will be good, we are quite sure that we can get around the need for anti-CD40. This is what allows us to see these huge primary immune responses, which may be very important for therapeutic vaccines, but for preventive vaccines the primary immune response may not be the best indicator for efficacy.

Ryffel: What could replace αGal-Cer?

Steinman: Poly IC by itself can give memory. Also it has been used in patients and is relatively cheap.

Finn: It has also been used as a single agent. You mentioned that there are both DEC-205 and Fcγ receptors. In terms of structural engineering of the anti-DEC-205, by putting a protein in the Fc portion you eliminate the ability of the Fc receptor binding. Perhaps it would be interesting to engineer DEC-205 such that some of the molecule goes to a DEC-205 receptor and some goes to the Fc receptor. You could activate the cell through two different pathways with a single agent then.

Steinman: I agree; it is a good idea.

Gordon: What species is the Fc in this construct?

Steinman: This is a mouse IgG1. It is a rat antibody in its variable (V) regions but the constant domains are mouse IgG1. It is then further mutated so that it doesn't bind Fc receptors.

Ryffel: The MyD88 dependence of the response is really intriguing.

Steinman: Yes. People are forgetting that these innate lymphocytes are really important but distinct ways to control the immune response.

Gordon: Is that true for the interferon pathways as well?

Steinman: We have looked at this and we can't find any type 1 interferon from the DCs.

Lambrecht: One of the receptors you mentioned when you discussed receptors for dying cells was the milk fat globulin/lactadherin pathway. Is there any progress in working out what ligands are involved? We have found these on exosomes from tumour cells.

Steinman: I am not aware of anyone who has shown that a specific receptor is working for the uptake of dying cells *in vivo*. *In vivo* experiments are what is missing.

Romani: Which cytokine drives maturation of DCs?

Steinman: The only cytokine that at this time by itself can be added to a DC and which will then allow the DC to induce a T cell response is TSLP, thymic stromal lymphopoietin. All the other cytokines are working together with something else. The interferons are the interesting ones, but they are typically given with GM-CSF.

Finn: The ability of the innate lymphocytes is largely untapped. Considering that we are talking about αGal-Cer as a reagent, there are some encouraging data from the work of the late Gordon Ross showing the recruitment of anti-tumour antibodies induced by vaccination and then when αGal-Cer is added the combination of the two has a very good anti-tumour effect in animal models.

Steinman: The problem has been to liberate the αGal-Cer for human studies. Many new glycolipid compounds have been made and these hopefully will soon be available for research in patients.

Reference

Casares N, Pequignot MO, Tesniere A et al 2005 Caspase-dependent immunogenicity of doxorubicin-induced tumor cell death. J Exp Med 202:1691–1701

Macrophage receptors and innate immunity: insights from dectin-1

Gordon D. Brown

Institute of Infectious Disease and Molecular Medicine, Faculty of Health Sciences, University of Cape Town, Observatory, 7925, Cape Town, South Africa

Abstract. The innate ability of macrophages to induce an immune response to pathogens is dependent upon germline encoded pattern recognition receptors which recognise conserved microbial structures. These receptors not only mediate pathogen recognition, but promote microbial uptake and killing and the induction of inflammatory responses. Although the recently described Toll-like receptors (TLR) have been shown to play a central role in mediating the intracellular signals involved, the non-TLRs also have important functions in these processes. Once such receptor is Dectin-1, a myeloid expressed signalling C-type lectin-like receptor which is involved in the innate recognition of fungal pathogens. Dectin-1 can induce a variety of cellular responses, including phagocytosis, the respiratory burst and cytokine production. These responses are mediated through novel signalling pathways induced from the cytoplasmic immuno-receptor tyrosine based activation-like motif of the receptor. Although the *in vivo* role of Dectin-1 has still to be fully elucidated, there is emerging evidence that this receptor plays a role in the inflammatory response to pulmonary fungal pathogens and that it is involved in certain autoimmune and respiratory diseases.

2006 Innate immunity to pulmonary infection. Wiley, Chichester (Novartis Foundation Symposium 279) p 114–126

The innate response is the first line of host defence against infection and is responsible for immediately recognizing and countering microbial invasion. The innate immune system is composed principally of phagocytic leukocytes, such as macrophages and neutrophils, which are responsible for ingesting and killing the invading organisms. How the innate immune system is able to recognize pathogens has been the focus of much interest and has led to the discovery of the evolutionarily ancient germ-line-encoded receptors, the pattern-recognition receptors (PRRs). These receptors recognise highly conserved microbial structures, the pathogen-associated molecular patterns (PAMPs), enabling the host to recognize a diverse range of pathogens quickly without the need for somatic recombination (Janeway 1992).

PRRs can be grouped into a number of families based on their structure and/or function, and their cellular location; either in the cytoplasm, membrane or serum/

TABLE 1 Pattern recognition receptors

Location	*Family*	*Selected examples*
Serum/Tissue Fluid	complement	C3, C1q
	lipid transferases	LBP
	collectins	SP-A, SP-D, MBL
	pentraxins	PTX3, SAP, CRP
Membrane	leucine-rich repeat proteins	TLRs, CD14
	scavenger receptors	SR-A, LOX-1, MARCO, CD36
	classical C-type lectins	mannose receptor, DC-SIGN, SIGNR1
	non-classical C-type lectins	Dectin-1
	integrins	CR3, CR4
	long PGRP	PGRP-L, PGRP-Iα, PGRP-Iβ
	other	lactosylceramide
Cytoplasmic	interferon-induced proteins	RNA-activated protein kinase, Mx protein GTPases
	NOD	NOD1, NOD2, NAIP

tissue fluids (Table 1). These receptors are used by leukocytes to directly recognize pathogens, a process termed non-opsonic recognition, which can also occur in the vacuole, after microbial uptake, or in the cytoplasm. Pathogen recognition can also be indirect, whereby distinct membrane receptors recognise serum or tissue fluid PRR-coated (opsonized) pathogens, in a process termed opsonic recognition. Some PRRs promote actin-dependent phagocytosis, following recognition, leading to microbial killing in the resultant phagosome, through various mechanisms.

PRRs also induce inflammatory responses to pathogens following recognition. These responses, which involve the production of cytokines, chemokines and other soluble factors, result in the activation and recruitment of other cells to the site of infection, leading finally to the initiation of the adaptive response. Although thought to be mediated primarily by the TLRs, there is growing evidence that the non-TLR PRRs contribute to this process, through the presentation of PAMPs, such as described for CD36 (Hoebe et al 2005) and CD14 (Jiang et al 2005), or by directly inducing intracellular signals, such as mediated by the C-type lectin-like receptor, Dectin-1. Thus PRRs not only mediate the recognition, uptake and killing of pathogens, but have a significant influence on the resultant immune response.

The TLR receptors

Originally identified in Drosophila, the TLR consists of a family of at least 11 proteins possessing extracellular leucine-rich repeat regions, which are involved in ligand recognition, and a conserved intracellular Toll/IL1R (TIR) domain which is involved in intracellular signal transduction. Despite similarity in their extracellular regions, the TLRs recognise a diverse, but receptor-specific, range of microbial structures. The TLRs also have different cellular locations: TLR1, 2 and 4 are expressed at the cell surface whereas TLR3, 7, 8 and 9 are located in intracellular compartments, including endosomes and the endoplasmic reticulum. It has been proposed that intracellular vacuoles are the main location for TLR recognition, even for the cell surface receptors, although there is evidence that these events can occur at the cell surface (Brown et al 2003). It is unclear how most TLRs recognize their ligands, although some interactions have been elucidated, such as the direct interaction of TLR5 with flagellin. Ligand recognition leads to TLR homo- or heterodimerization and the initiation of specific signalling cascades mediated through intracellular TIR-containing adaptors, including MyD88, TIRAP (MAL), TRIF (TICAM1) and TRAM (TICAM2) (Akira & Takeda 2004). These cascades, which are still incompletely understood, result in the activation of several transcription factors, such as NF-κB and IRF3, inducing the production of proinflammatory cytokines and chemokines as well as TLR-specific patterns of gene expression. While it is clear how the activation of TLRs can lead to an inflammatory Th1-type response, the mechanisms behind the generation of Th2 responses are still unclear, although the TLRs have been implicated. The specificity of these responses is also incompletely understood but is thought to stem, at least in part, from the association with a particular adaptor(s), heterodimerization (such in the case of TLR2 with TLR1 or TLR6), and the contribution of other non-TLR PRRs, such as Dectin-1, which are associated with the recognition of specific microbes. Here we will discuss Dectin-1 as model non-TLR PRR, highlighting its contribution to these inflammatory responses and its role in pulmonary infection and disease.

Dectin-1

Structure, expression and function

Dectin-1 is a non-classical, or natural killer (NK) like, C type lectin receptor with a single extracellular carbohydrate-recognition domain (CRD) on a stalk, a transmembrane region, and a cytoplasmic tail containing an immunoreceptor tyrosine based activation (ITAM)-like motif (Fig. 1) (Ariizumi et al 2000). Human, and possibly murine, Dectin-1 is alternatively spliced into a number of smaller isoforms, the most predominant of which lacks the stalk region. These isoforms are expressed in a cell-specific manner and, although they appear to function similarly *in vitro* (Willment et al 2001), they may possess different activities, as shown for the isoforms of other C-type lectins.

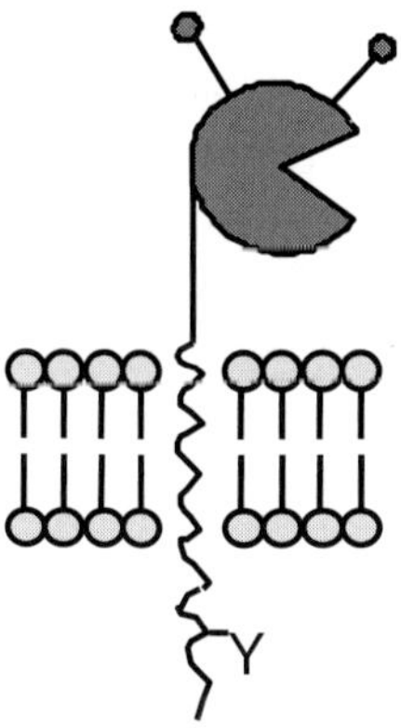

FIG. 1. Cartoon structure of murine Dectin-1. Dectin-1 is a type II transmembrane NK-like C type lectin possessing a carbohydrate recognition domain, a stalk and transmembrane region, and a cytoplasmic tail containing an immunoreceptor tyrosine based activation (ITAM)-like motif (the non-typical YxxL motif is underlined).

Dectin-1 is expressed predominantly by myeloid cells, including macrophages, monocytes, dendritic cells and neutrophils, as well as on a subset of T cells (Taylor et al 2002), but shows heterogeneous levels of expression on these cells in tissues (Reid et al 2004). In the lung, in particular, and other portals of pathogen entry, Dectin-1 is highly expressed (Taylor et al 2002, Reid et al 2004), consistent with a role in immune surveillance. This receptor is also expressed by B cells and eosinophils in humans (Willment et al 2005), but not in mice, although the functional significance of this species difference is unclear. The levels of Dectin-1 expression can also be modulated by cytokines, steroids and microbial stimuli (Willment et al 2003).

Dectin-1 was identified as a receptor for β-glucans from a functional screen of a macrophage cDNA library (Brown & Gordon 2001). Dectin-1 recognizes soluble and particulate β(1→3)- and /or β(1→6)-linked glucans, including zymosan, a β-glucan rich cell wall extract of *Saccharomyces cerevisiae* widely used as a stimulatory particle to study immune function *in vitro* and *in vivo*. These carbohydrates are found primarily in fungal cell walls, but also in plants and some bacteria. The ability of Dectin-1 to recognise carbohydrates is unusual as the receptor lacks the conserved residues that are required for carbohydrate binding in the classical C-type lectins. Although it is still unclear how Dectin-1 recognises β-glucans, two residues, Trp221 and His223, have been shown to be critical for β-glucan binding (Adachi et al 2004). The use of specific receptor antagonists, blocking monoclonal antibodies and cells from knockout mice has demonstrated that Dectin-1 is the major β-glucan receptor on leukocytes (Brown et al 2002, Gantner et al 2003, Rogers et al 2005, Willment et al 2005, Taylor et al, unpublished data 2005).

Dectin-1 was originally identified as a receptor recognizing T-cells (Ariizumi et al 2000) and may be able to stimulate T cell proliferation (Ariizumi et al 2000, Grunebach et al 2002), suggesting that the receptor may act as a co-stimulatory molecule. Dectin-1 is expressed on macrophages and dendritic cells in the T cell areas of the spleen, lymph nodes and thymus (Reid et al 2004), supporting a role for this receptor in the interactions between antigen-presenting cells and T cells. However, the T-cell ligand is unknown and is probably not a carbohydrate (T. Feizi et al unpublished data 2005, Brown & Gordon 2001, Willment et al 2001), indicating that Dectin-1 may have two separate ligand binding sites, one for the endogenous T cell ligand and one for β-glucans.

Dectin-1 mediated signalling and cellular responses

Dectin-1 can induce its own intracellular signals upon β-glucan or fungal recognition, leading to a variety of cellular responses. This is mediated by the ITAM-like motif in the cytoplasmic tail of the receptor, which becomes tyrosine phosphorylated upon ligand binding (Gantner et al 2003). The motif differs slightly from the tandem YxxL motif found in traditional ITAM sequences of other activation receptors (see Fig. 1), and only the membrane proximal YxxL repeat appears to be necessary for Dectin-1-mediated signalling. As a signalling receptor, Dectin-1 is the first example of a non-TLR PRR playing a direct role in the cellular response to pathogens.

Dectin-1-mediated signalling is complex and involves multiple pathways which are mostly undefined (Fig. 2). In response to particulate β-glucan ligands Dectin-1-mediated signalling induces phagocytosis and the production of proinflammatory cytokines and chemokines, including TNFα, IL12 and MIP2, although the latter function requires a collaborative signalling with TLR2 and TLR6 (Brown et al 2003, Gantner et al 2003, Steele et al 2003, Herre et al 2004). This was the first description of a non-TLR PRR being directly involved in the signal transduction leading to an inflammatory response and the first demonstration of such a collaboration with the TLRs. Dectin-1 can also signal via a novel interaction with Syk kinase, in a cell-specific manner, to induce the respiratory burst and the production of IL10 and IL2 (Rogers et al 2005, Underhill et al 2005). Overall, these signalling mechanisms of Dectin-1 may be representative of other PRRs which possess similar cytoplasmic sequences. Indeed, a related C-type lectin, CLEC-2, was recently shown to activate platelets via Syk, through a similar cytoplasmic motif (Suzuki-Inoue et al 2005).

Dectin-1 and its role in pulmonary infection and disease

There has been a significant increase in fungal infections over the last few decades which now represent over 10% of all nosocomial infections; a result of the increase

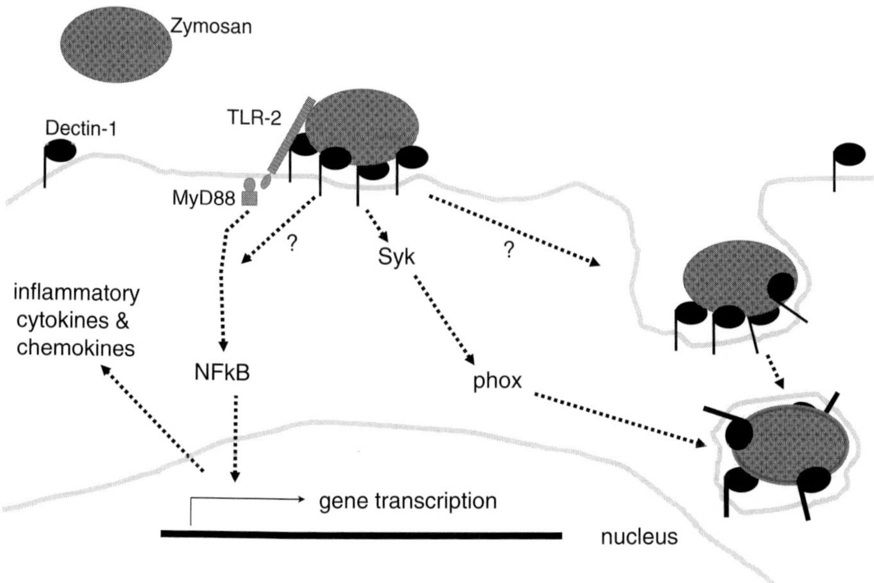

FIG. 2. Dectin-1-mediated intracellular signalling in macrophages. Upon binding of β-glucan bearing ligands, such as zymosan, Dectin-1 induces a number of intracellular signals through mostly unknown pathways. These signals induce microbial uptake through phagocytosis and the inflammatory response, in collaboration with TLR2/6. Dectin-1 is able to signal via Syk kinase, in a novel fashion, inducing the respiratory burst. The involvement of Syk is cell-specific and signalling via this kinase is different in dendritic cells, resulting in IL10 and IL2 production (not shown).

in AIDS and use of immunosuppressive treatments, such as transplantation-rejection therapy (Romani 2004). In healthy hosts, protection against fungal infection requires a Th1-type immune response which induces the activation of phagocytes and their fungicidal activities, such as the respiratory burst. The timely induction of these protective mechanisms is reliant upon the innate recognition of the pathogens, which occurs through several opsonic and non-opsonic receptors, including a number of TLRs and Dectin-1 (Table 2).

Dectin-1 can mediate the cellular recognition of a number of fungal pathogens, including *Coccidoides spp.* (Viriyakosol et al 2005), *Candida* spp. (Brown et al 2003), *Pneumocystis* spp. (Steele et al 2003), *Aspergillus* spp. (Steele et al 2005) and *Saccharomyces* spp. (Brown et al 2003) (Fig. 3). This recognition is mediated though exposed fungal β-glucans, which can represent more than 50% of the cell wall, but may be restricted to specific surface regions, such as the bud scar, in certain fungal species (Gantner et al 2005). The interaction with Dectin-1 also leads to fungal uptake and killing, through the induction of the respiratory burst (Steele et al 2003, Gantner et al 2005), and the production of protective cytokines and chemokines,

TABLE 2 Pattern recognition receptors (PRRs) involved in fungal recognition

PRR	*Selected fungal PAMPs*
C3	Fungal surfaces
CD14	Glucuronoxylomannan
Chitinase	Chitin
Chitotriosidase	Chitin
CR3	Mannose, β-glucan, N-acetylglucosamine, complement opsonised pathogens
DC-SIGN	internal mannose, terminal di-mannose
Dectin-1	β-glucan
Lactosylceramide	β-glucan
Mannose receptor	Terminal mannose
Mannose-binding lectin	Mannose, fucose, glucose
Pentraxin 3	Galactomannan, zymosan
SP-A	Mannose, fucose, glucose
SP-D	Mannose, glucose, β-glucan
TLR2	Phospholipomannan, zymosan, lipoproteins, lipopetides, glycolipids
TLR4	Mannan, glucuronoxylomannan
TLR9	CpG DNA

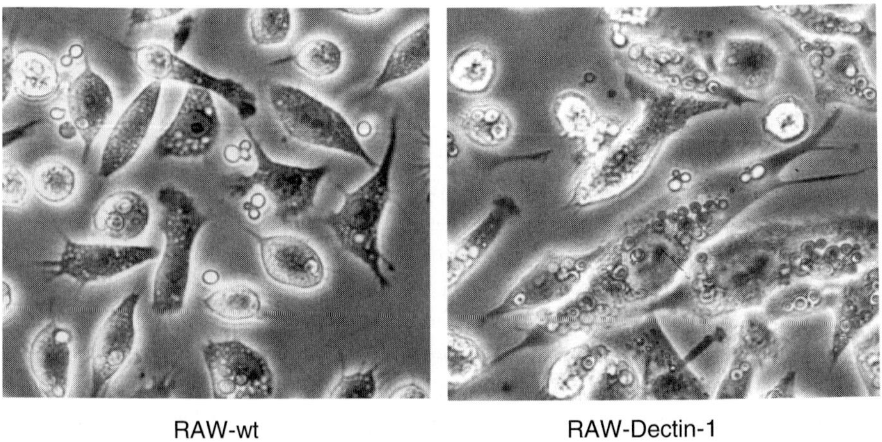

FIG. 3. Dectin-1 mediates recognition of live fungal particles. Through its ability to bind surface exposed β-glucans, Dectin-1 plays an important role in the innate recognition and response to fungal particles. Here, the over expression of Dectin-1 in RAW macrophages (RAW-Dectin-1) can be seen to confer a greatly enhanced ability to bind and ingest live *Saccharomyces cerevisae*, when compared to control (RAW-wt) cells.

such as TNFα (Brown et al 2003) and MIP2 (Steele et al 2003, Viriyakosol et al 2005). Dectin-1 also induces the production of IL12, a cytokine critical for inducing a Th1 response. Thus Dectin-1 appears to have a central role in inducing protective immune responses to fungal pathogens.

Emerging *in vivo* evidence also supports a role for Dectin-1 as a key receptor in anti-fungal immunity. By far the most compelling data are from Dectin-1 deficient mice which show a greatly increased susceptibility to systemic infection with *Candida albicans*, but are otherwise apparently normal (P. Taylor, M. Botto, K. Haynes, S. Gordon and G. D. Brown, unpublished data). In addition, blockage of Dectin-1 inhibited inflammatory cytokine production and cellular recruitment in the lungs of wild-type mice infected with *Aspergillus fumigatus*, highlighting the central role of Dectin-1 in the recognition of this pulmonary pathogen (Steele et al 2005). This inflammatory response was largely TLR2 independent in alveolar macrophages, indicating that Dectin-1 may be capable of directly inducing inflammatory cytokine production in these cells.

While recognition of β-glucans appears to be important for the establishment of a successful immune response, fungal pathogens may limit the recognition of these carbohydrates. For some pathogens, such as *Candida albicans* (Gantner et al 2005) and *Aspergillus fumigatus* (Steele et al 2005), protective immune responses are only induced to specific morphological forms of these organisms, when the β-glucans are exposed. In contrast to the yeast form, for example, *Candida* hyphae lack surface exposed β-glucans and do not induce protective responses. In other species, β-glucans are masked by encapsulation, as occurs in *Cryptococcus neoformans* (Cross & Bancroft 1995), or the cell wall β-glucan content is changed upon infection, as occurs in *Paracoccidioides brasiliensis* (Borges-Walmsley et al 2002). This suggests that the avoidance of Dectin-1-mediated recognition may have evolved in these pathogens as a means of subverting anti-fungal immunity.

Fungal components, such as β-glucans, are also thought to be involved in a variety of human diseases, including respiratory disorders such as asthma. *Aspergillus fumigatus*, for example, can induce allergic bronchopulmonary aspergillosis (ABPA), hypersensitivity pneumonitis and allergic asthma in non-immunocompromised individuals. As Dectin-1 can induce β-glucan dependent pulmonary inflammation in response to this pathogen (Steele et al 2005), it is likely to play a role in the generation of these diseases. In addition, in genetically susceptible mice, Dectin-1 is able to induce autoimmune arthritis after exposure to fungi or after the direct administration of β-glucans, although the mechanisms leading to this disease are unclear (Yoshitomi et al 2005). Overall, these data demonstrate that proinflammatory signalling induced by Dectin-1, while required for protective responses to fungal pathogens, can also lead to autoimmunity and disease.

Summary

In summary, the study of Dectin-1 has shown us that:

(1) Non-TLR PRRs play a significant role in the initiation of immune responses to pathogens, both through the presentation of PAMPs to TLRs and through the induction of intracellular signals.
(2) Non-TLR PRRs contribute to important aspects of anti-microbial host-defence, including microbial uptake and killing.
(3) Non-TLR PRRs can utilize novel and mostly undefined pathways to induce intracellular signals.
(4) Non-TLR PRRs can contribute to autoimmunity and disease.

Acknowledgements

I am grateful to the Wellcome Trust, Edward Jenner Institute for Vaccine research and the National Institutes of Health for financial support. I thank J. Willment for proof reading this manuscript. GDB is a Wellcome Trust Senior Research Fellow in Biomedical Science in South Africa.

References

Adachi Y, Ishii T, Ikeda Y et al 2004 Characterization of beta-glucan recognition site on C-Type lectin, dectin 1. Infect Immun 72:4159–4171
Akira S, Takeda K 2004 Toll-like receptor signalling. Nat Rev Immunol 4:499–511
Ariizumi K, Shen GL, Shikano S et al 2000 Identification of a novel, dendritic cell-associated molecule, dectin-1, by subtractive cDNA cloning. J Biol Chem 275:20157–20167
Borges-Walmsley MI, Chen D, Shu X, Walmsley AR 2002 The pathobiology of Paracoccidioides brasiliensis. Trends Microbiol 10:80–87
Brown GD, Gordon S 2001 Immune recognition: A new receptor for beta-glucans. Nature 413: 36–37
Brown GD, Taylor PR, Reid DM et al 2002 Dectin-1 is a major beta-glucan receptor on macrophages. J Exp Med 296:407–412
Brown GD, Herre J, Williams DL et al 2003 Dectin-1 mediates the biological effects of beta-glucan. J Exp Med 197:1119–1124
Cross CE, Bancroft GJ 1995 Ingestion of acapsular Cryptococcus neoformans occurs via mannose and beta-glucan receptors, resulting in cytokine production and increased phagocytosis of the encapsulated form. Infect Immun 63:2604–2611
Gantner BN, Simmons RM, Canavera SJ, Akira S, Underhill DM 2003 Collaborative induction of inflammatory responses by dectin-1 and toll-like receptor 2. J Exp Med 197:1107–1117
Gantner BN, Simmons RM, Underhill DM 2005 Dectin-1 mediates macrophage recognition of Candida albicans yeast but not filaments. EMBO (Eur Mol Biol Organ) J 24:1277–1286
Grunebach F, Weck MM, Reichert J, Brossart P 2002 Molecular and functional characterization of human Dectin-1. Exp Hematol 30:1309–1315
Herre J, Marshall AJ, Caron E et al 2004 Dectin-1 utilizes novel mechanisms for yeast phagocytosis in macrophages. Blood 104:4038–4045
Hoebe K, Georgel P, Rutschmann S et al 2005 CD36 is a sensor of diacylglycerides. Nature 433:523–527

Janeway CA Jr 1992 The immune system evolved to discriminate infectious nonself from non-infectious self. Immunol Today 13:11–16 Review
Jiang Z, Georgel P, Du X et al 2005 CD14 is required for MyD88-independent LPS signaling. Nat Immunol 6: 565–570
Reid DM, Montoya M, Taylor PR et al 2004 Expression of the beta-glucan receptor, dectin-1, on murine leukocytes in situ correlates with its function in pathogen recognition and reveals potential roles in leukocyte interactions. J Leukoc Biol 76:86–94
Rogers NC, Slack EC, Edwards AD et al 2005 Syk-dependent cytokine induction by dectin-1 reveals a novel pattern recognition pathway for C-type lectins. Immunity 22:507–517
Romani L 2004 Immunity to fungal infections. Nat Rev Immunol 4:1–23
Steele C, Marrero L, Swain S et al 2003 Alveolar macrophage-mediated killing of Pneumocystis carinii f. sp. muris involves molecular recognition by the dectin-1 β-glucan recptor. J Exp Med 198:1677–1688
Steele C, Rapaka R, Metz A et al 2005 The beta glucan receptor dectin-1 recognizes specific morphologies of Aspergillus fumigatus. PLOS pathogens, 1:e42
Suzuki-Inoue K, Fuller GL, Garcia A et al 2005 A novel Syk-dependent mechanism of platelet activation by the C-type lectin receptor CLEC-2. Blood 107:542–549
Taylor PR, Brown GD, Reid DM et al 2002 The beta-glucan receptor, dectin-1, is predominantly expressed on the surface of cells of the monocyte/macrophage and neutrophil lineages. J Immunol 269:3876–3882
Underhill DM, Rossnagle E, Lowell CA, Simmons RM 2005 Dectin-1 activates Syk tyrosine kinase in a dynamic subset of macrophages for reactive oxygen production. Blood 106:2543–2550
Viriyakosol S, Fierer J, Brown GD, Kirkland TN 2005 Innate immunity to the pathogenic fungus Coccidioides posadasii is dependent on toll-like receptor 2 and dectin-1. Infect Immun 73:1553–1560
Willment JA, Gordon S, Brown GD 2001 Characterisation of the human beta-glucan receptor and its alternatively spliced isoforms. J Biol Chem 276:43818–43823
Willment JA, Lin HH, Reid DM et al 2003 Dectin-1 expression and function is enhanced on alternatively activated and GM-CSF treated macrophages and negatively regulated by IL-10, dexamethasone and LPS. J Immunol 171:4569–4573
Willment JA, Marshall AS, Reid DM et al 2005 The human beta-glucan receptor is widely expressed and functionally equivalent to murine Dectin-1 on primary cells. Eur J Immunol 35:1539–1547
Yoshitomi H, Sakaguchi N, Kobayashi K et al 2005 A role for fungal β-glucans and their receptor Dectin-1 in the induction of autoimmune arthritis in genetically susceptible mice. J Exp Med 201:949–960

DISCUSSION

R Sim: In contrast with some of the pattern-recognition receptors (PRRs) we have heard about so far, Dectin-1 seems to have only a single binding domain. Unlike some of the others, it therefore doesn't rely on multiple low-affinity interactions. If you make recombinant material, will the monomer of Dectin-1 still bind to the ligand?

Brown: We have done some vital studies in collaboration with David Williams (East Tennessee State University), which showed that it is a high affinity interaction. He says it is 10^{-15} M, which is very high.

Peiris: Is there any evidence that viruses bind to Dectin-1 and enter cells through this pathway?

Brown: Not yet. I have to be cautious about this because Dectin-1 has T cell ligands. But our current knowledge would suggest that this receptor does not bind viruses.

Peiris: You highlighted that Toll-like receptor 2 (TLR2) was cooperating with Dectin-1. Have you done TLR2 knockouts?

Brown: Yes, it is required for inflammatory responses.

Segal: Those macrophages from the Dectin-1 knockout look different from normal ones.

Brown: Those are not cells from a knockout, but from a cell line. The reason they look different is because there is zymosan in the medium, and the cell is responding to this.

Segal: One had microfilia and the other one didn't.

Brown: That's an artefact of the system.

McGreal: In the experiments where cells are transfected with Dectin-1 that has been engineered without its cytoplasmic tail or with mutated forms of the ITIM motif you clearly see blunted responses to yeast and yeast derived particles. Do you also see a defect in the recruitment of TLR2?

Effective signalling in response to Dectin-1 ligands requires colocalization of Dectin 1 and TLR2, is it possible therefore that blunted responses in the Dectin-1 mutants is partially a result of a physical dislocation of these two receptors as a result of the mutated Dectin-1 cytoplasmic tail?

Brown: We haven't looked at that. Dectin and TLR2 certainly colocalize in the phacocytic cup. I would expect there would still be the same recruitment.

Quesniaux: TLR2 and 4 seem to be involved in mycobacteria internalization, since absence of both TLR2 and TLR4 leads to 36% uptake inhibition by macrophages (Nicolle et al 2004a). This inhibition is more accentuated in the absence of MyD88 (Nicolle et al 2004b). Is there any chance that Dectin-1 interaction with TLR2 is involved in that process?

Brown: A recent paper by David Russell's group has shown that this isn't the case. There is evidence to suggest that they modulate the phagocytic process, but they are not directly involved in phagocytosis.

E Sim: Can you give some indication of the level of Dectin-1 expression in your transducers, compared with what you find normally in the RAW cells?

Brown: In the RAW cells there are low levels of endogenous receptor, which is markedly increased (greater than 10-fold) in the transductants.

Mantovani: You emphasized differences between macrophages and DCs. In my mind, this raises the question, how faithful are these couples? In other words, in different cellular contexts would Dectin-1 and TLR2 behave differently?

Brown: That is a good question. We are looking at this now. We have the defined ligands of Dectin and Toll, and are looking at different cells to see how faithful their coupling is.

Romani: Because Dectin-1 seems to be equally active against *Aspergillus* and *Candida*, can you tell us about the relative distribution of Dectin-1 *in vivo* in the lung? Also, we need to explain why TLR2-deficient mice have different susceptibility to *Candida* and *Aspergillus*.

Brown: What we have found with *Aspergillus* and *Candida* is that the recognition of the fungus is stage-specific. In *Candida* the yeast form is recognized by Dectin-1 whereas hyphae are not. In *Aspergillus*, conidia are recognized by Dectin-1 but hyphae are not. Different forms seem to trigger different inflammatory responses. In mice, in the lung, Dectin-1 is very highly expressed in alveolar macrophages. My collaborator, Dr Chad Steele (University of Pittsburgh), has taken alveolar macrophages and looked at them in inflammatory responses to *Aspergillus*. He can block almost every inflammatory cytokine by blocking Dectin-1 function. He also has evidence suggesting that some of the cytokines are TLR2 independent.

Romani: Is it also happening in the stomach?

Brown: No, we haven't found it there. It is expressed in the intestine though.

Segal: How specific is it for the yeasts? Does it bind any other sorts of microbes?

Brown: It seems to be highly specific for β1,3-glucans. There are bacterial species that produce polysaccharide ligands. At the moment we know that it is some bacteria and fungi.

Latgé: It only binds fibrillar β1,3-glucans, and not the soluble form of β1,3-glucans. These data have been obtained with a recombinant Dectin-1 using an ELISA inhibition format and various β1,3-glucans of different sizes.

Brown: We have found that the longer chain ones give much better inhibition.

Romani: It is known that β-glucan stays inside the cell wall and it is somehow inhibited by the external fibrillar portion of the cell wall of *Candida*. How can you envision Dectin-1 picking up the β-glucan that is on the inside of the wall?

Brown: The cell wall structure of yeast is wrong! It appears to be that in some areas β-glucans are exposed to the surface. There are two studies that should be mentioned here. One was done by Underhill's group which shows that in *Candida*, the bud scar is highly enriched in β-glucan (Gantner et al 2005). There is another paper by Antonio Cassone who has made a β-glucan vaccine to fungi (Torosantucci et al 2005). They used an anti-β1,3-glucan antibody and could also detect this carbohydrate on the surface.

Latgé: We have shown that the bud scar contains fibrillar β1,3-glucan. The anti-β1,3-glucan antibody used for vaccine give a different labelling than Dectin-1.

Speert: There is a condition called chronic mucocutaneous candidiasis. Has anyone looked to see whether this involves a Dectin-1 deficiency?

Brown: Desa Lilic (University of Newcastle) has looked at a couple of patients with this. At that stage no antibody was available, so she looked at mRNA only and found no deficiency.

Finn: I want to go back to T cells. Jonathan Sprent has published some papers on T cells eating other cells (Sprent 2005), taking chunks of cells and swallowing them up. As far as I know the receptor involved has not been identified. Similarly, Simon Barratt-Boyes has shown that DCs eat other cells, including DCs (Harshyne et al 2003). Perhaps there is a specific ligand involved in this process.

Brown: We have looked for endogenous ligands but with no success.

Finn: It is very interesting that just binding without endocytosis would activate a response. It would be especially helpful with something like neutrophils, if they can kill without needing to take anything in. Is it expressed on neutrophils?

Brown: Yes, two weeks ago we did our first neutrophil experiments with the knockout and wild-type enzymes. There was no binding of zymosan, which is not what we expected, and there was a considerably reduced respiratory burst in that population of cells.

References

Gantner BN, Simmons RM, Underhill DM 2005 Dectin-1 mediates macrophage recognition of Candida albicans yeast but not filaments. EMBO J 24:1277–1286

Harshyne LA, Zimmer MI, Watkins SC, Barratt-Boyes SM 2003 A role for class A scavenger receptor in dendritic cell nibbling from live cells. J Immunol 170:2302–2309

Nicolle D, Fremond C, Pichon X, Bouchot A, Maillet I, Ryffel B, Quesniaux VJF 2004a Long-term control of Mycobacterium bovis BCG infection in the absence of Toll-like receptors: Investigation on TLR2, TLR6 or TLR2-TLR4 deficient mice infection immunity 72: 6994–7004

Nicolle D, Pichon X, Bouchot A et al 2004b Chronic pneumonia despite adaptive immune response to Mycobacterium bovis BCG in MyD88-deficient mice. Lab Invest 84:1305–1321

Sprent J 2005 Swapping molecules during cell-cell interactions Sci STKE. 273:pe8

Torosantucci A, Bromuro C, Chiani P et al 2005 A novel glyco-conjugate vaccine against fungal pathogens. J Exp Med 202:597–606

Toll-like receptors and control of mycobacterial infection in mice

Bernhard Ryffel*†, Muazzam Jacobs†, Shreemanta Parida‡, Tania Botha§, Dieudonnée Togbe* and Valerie Quesniaux*

** CNRS, UMRG218 Orleans, France, †Infectious Disease Institute, University of Cape Town, Cape Town, South Africa, ‡Armauer Hansen Research Institute, Addis Ababa, Ethiopia and §Cape Technikon, Cape Town, South Africa*

Abstract. Microbial products including mycobacterial antigens are recognized by distinct Toll-like receptors (TLRs) resulting in activation of cells of the innate immune system. Ablation of most of the TLR signalling in mice deficient for the common adaptor protein MyD88 revealed that TLRs are crucial for the activation of an innate immune response as MyD88-deficient mice are highly sensitive to infection with *Mycobacterium tuberculosis.* Despite the profound defect of the innate immune response, MyD88 deficiency allows the emergence of an adaptive immunity. These data demonstrate that activation of multiple TLRs contributes to an efficient innate response to mycobacteria, while MyD88-dependent signalling is dispensable to generate adaptive immunity.

2006 Innate immunity to pulmonary infection. Wiley, Chichester (Novartis Foundation Symposium 279) p 127–141

Infectious tuberculosis due to *Mycobacterium tuberculosis* (*Mtb*), an intracellular pathogen capable of surviving and persisting within host mononuclear cells, represents a major global health challenge. A coordinated response of cells of the innate and adaptive immune system is required to control infection (Flynn & Chan 2001a, North & Jung 2004). Pathogen sequestration in macrophages within granulomas, dynamic structures containing activated lymphocytes and macrophages, contains the spread of infection. Gene targeted mice and neutralizing antibodies demonstrated critical mediators controlling *Mtb* infection, including interferon (IFN)γ, interleukin (IL)12, IL23, tumour necrosis factor (TNF), lymphotoxins, CD40 and nitric oxide (Flynn 2004, Cooper et al 2002, Ehlers et al 2003, Holscher et al 2001, Lazarevic et al 2003, Roach et al 2001, 1999, Garcia et al 2000, Jacobs et al 2000b). Mycobacteria have evolved to resist the eradication by macrophages by using elaborate evasion mechanisms (Flynn & Chan 2003). In a healthy host, a subclinical latent or chronic infection may persist (Gomez & McKinney 2004). Neutraliztion of IFNγ or TNF, inhibition of inducible nitric oxide synthase (iNOS), or

T cell depletion leads to reactivation of latent infection (Chan et al 1995, Flynn et al 1998, Flynn & Chan 2001b, Mohan et al 2001, Scanga et al 1999, 2000, Botha & Ryffel 2003).

Persistent macrophage and T lymphocyte activation control the viable, but sequestered bacilli within phagocytes in the granuloma structure. We hypothesize that mycobacterial products released by the sequestered bacilli, such as glycolipids, lipomannan (LM), phosphatidyl-myo-inositol mannoside (PIM), lipoarabinomannan (LAM), lipoproteins and other mycobacterial factors, may contribute to continued macrophage and dendritic cell activation through pathogen pattern recognition receptors (PRRs) such as Toll-like receptors (TLRs) and others.

Mammalian TLRs represent a structurally conserved family of membrane receptors, which have homology to the *Drosophila* Toll system (Medzhitov et al 1997, Akira & Takeda 2004). Microbial products activate mammalian TLRs inducing gene transcription regulating the adaptive immune response, resulting in chemokines, cytokines and costimulatory molecules. The TLR family now consists of 11 members, with TLR11 being discovered to be critical in the control of uropathogenic bacteria (Zhang et al 2004, Quesniaux & Ryffel 2004). The greatest variety of TLR mRNAs is found in professional phagocytes, suggesting a key role of TLRs in innate immunity. The main ligands for the individual TLRs are shown in Fig. 1.

TLRs are type I membrane proteins containing an extracellular domain with 19–25 tandem copies of leucine-rich repeats (LRRs) and a cytoplasmic Toll/IL1 receptor (TIR) domain similar to that of the IL1 receptor family (Takeda & Akira 2004, Akira & Takeda 2004). TLR signal transduction is mediated by binding of the adaptor protein MyD88 to the TIR domain of TLRs, followed by the recruitment of IL1 receptor associated kinases (IRAKs), TNF receptor associated factor (TRAF) 6, TGFβ-activated protein kinase 1 (TAK1), mitogen-activated protein (MAP) kinase and NF-κB activation (Akira et al 2003, Takeda & Akira 2004). Mal/TIRAP participates in signalling of TLR2 and TLR4, together with MyD88 (Horng et al 2002, Yamamoto et al 2002a, O'Neill et al 2003). TIR domain-containing adaptor inducing IFNβ (TRIF), also known as TIR-containing adaptor molecule 1 (TICAM1) or Lps2, has been identified (Yamamoto et al 2003, 2002b, Oshiumi et al 2003). TRIF is particularly important for interferon regulatory factor 3 (IRF3) activation mediated by viral-induced TLR3 engagement, but it is also involved in the TLR4 MyD88-independent activation of costimulatory molecules CD40 and CD86, through an IFNβ autocrine/paracrine loop (Hoebe et al 2003, Yamamoto et al 2003, 2002b). Pathogen binding to specific TLRs or to combinations of TLRs may recruit different adaptor proteins allowing a specific signalling cascade and gene activation programmes. The serine/threonine kinase family consists of two active kinases, IRAK and IRAK4, and two inactive or inhibitory kinases, IRAK2 and IRAKM (also named IRAK3).

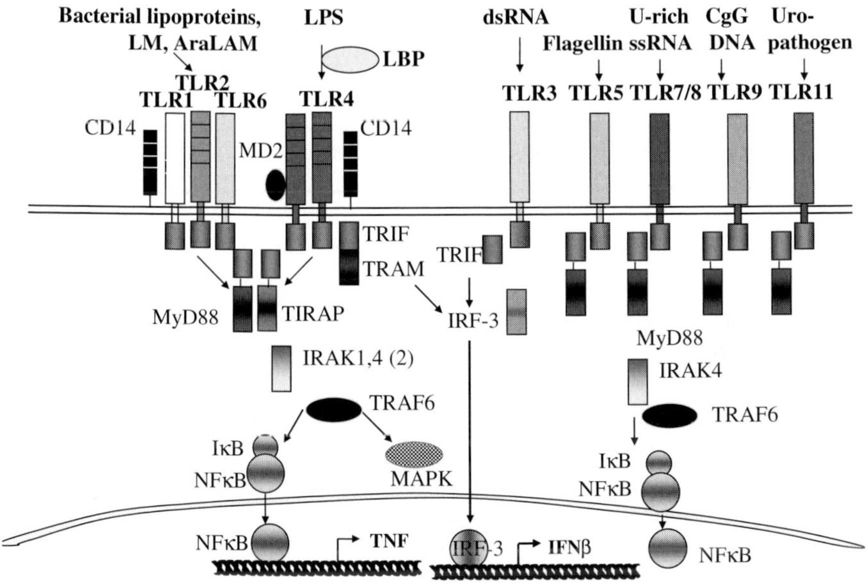

FIG. 1. Microbial ligands and association with known TLRs and adaptor molecules. Schematic representation of the structure of TLRs and the major TLR ligands. Most TLRs form homodimers, while TLR2 associates with either TLR1 or TLR6. TLR signalling is mediated through adaptors such as MyD88, TIRAP, TRIF or TRAM (Reproduced from Ryffel et al 2005, with permission).

This review is aimed at discussing current knowledge about the interaction of mycobacterial ligands or mycobacteria such as *Mtb* with TLRs and their role in controlling mycobacterial infection in gene deficient mice.

Mycobacterial TLR ligands and responses *in vitro*

The mycobacteria cell wall is composed of different glyocolipids such as LAM, mycolic acid, lipopeptides and phosphoinositol, which may be recognized by the immune system (Chatterjee & Khoo 1998, Daffe & Draper 1998, Nigou et al 2002). So far, TLR2, TLR4 and TLR1/TLR6 that heterodimerize with TLR2, have been implicated in the recognition of mycobacterial antigens (Bulut et al 2001, Hajjar et al 2001).

TLR2-dependent cell activation has been described for LAM from rapidly growing mycobacteria, lipomannan (LM), PIM (phosphatidyl-*myo*-inositol mannoside), or the 19 kDa mycobacterial lipoprotein (Aliprantis et al 1999, Brightbill et al 1999, Means et al 1999a, Jones et al 2001b, Gilleron et al 2003, Barnes et al 1992, Vignal et al 2003, Quesniaux et al 2004b). LMs, the biosynthetic precursors of LAMs, represent another class of abundant pro-inflammatory molecules of the mycobacterial cell wall. We showed that LMs from various mycobacterial origins are potent activators of proinflammatory cytokines in macrophages requiring TLR2 signalling (Quesniaux et al 2004b). LM induces CD40 and CD86 cell surface expression, abundant cytokine expression and nitric oxide production (Vignal et al 2003, Quesniaux et al 2004b). PIMs, the anchor motifs of LM and LAM, have pro-inflammatory activities (Jones et al 2001a). Dimannoside (PIM_2) and hexamannoside (PIM_6), the two most abundant classes of PIM found in *M. bovis* BCG and *Mtb* H37Rv, were recently shown to activate macrophages to secrete TNF through TLR2, irrespective of their acylation pattern, and to signal through MyD88 (Gilleron et al 2003). Based on the present knowledge the balance between PIM, LM and LAM synthesis by pathogenic mycobacteria might provide pro- or anti-inflammatory immunomodulatory signals during primary infection, but also during latent infection.

Mycobacterial lipoproteins were shown to activate antigen-presenting cells (APC) through TLR2 signalling (Underhill et al 1999, Brightbill et al 1999). Recent reports further suggest that the 19 kDa lipoprotein, LpqH (Rv3763) has also TLR2-dependent inhibitory functions on IFNγ regulated responses, including MHC class II antigen processing in macrophages (Noss et al 2000, 2001, Pai et al 2003, Gehring et al 2003). Further, the 24 kDa lipoprotein, LprG (Rv1411c) also appears to inhibit in MHC-II antigen processing and hence $CD4^+$ T cell activation, although short-exposure induces TLR2-dependent TNF production (Gehring et al 2004). Furthermore, a phenolic glycolipid (PGL) from a virulent *Mtb* strain has been shown to inhibit innate immune responses (Reed et al 2004). A more detailed discussion of mycobacterial ligands, their receptor specificity and biological properties is given elsewhere (Quesniaux et al 2004b, 2004a).

Viable and killed *Mtb* bacilli (virulent and attenuated) activate CHO cells and murine macrophages that express either TLR2 or TLR4 (Means et al 1999b). Macrophages expressing a dominant-negative mutant for MyD88 failed to react to mycobacteria, underlining the requirement of TLRs mediating the downstream signalling cascade responsible for the transcription of TNF. Using bone marrow-derived macrophages derived from TLR2 and/or TLR4 deficient mice, we showed TLR2- and to a lesser extent TLR4-dependent activation of TNF and IL12 production after infection with live *M. bovis* BCG (Fremond et al 2003, Nicolle et al 2004b), which is completely abrogated in MyD88 deficient macrophages (Nicolle et al 2004a). However, recognition of heat-killed *M. bovis* BCG, extensive-freeze-dried *M. bovis* BCG, a soluble fraction of *M. bovis* BCG culture supernatant, or a 'well dispersed' live *M. bovis* BCG cultivated in the presence of detergent, was predominantly mediated through TLR2, as essentially no response remained in TLR2-deficient macrophages or dendritic cells (Nicolle et al 2004b). Neither TLR1 nor TLR6 signalling on their own are critical for mycobacteria macrophage activation induced by these mycobacterial preparations as TLR1- or TLR6-deficient macrophages respond normally suggesting a potential compensation of TLR1/6 (unpublished). Further, mycobacteria induced nitric oxide production, a potent antimycobacterial effector molecule, in primary macrophages in a TLR2-dependent way, in contrast with previous reports (Means et al 2001). Recent work demonstrated that MyD88 is crucial for macrophages to acquire a normal IFNγ response (Ehrt et al 2001). Interestingly, the expression of CD40, CD80 and CD86 on macrophages and dendritic cells was not affected by the absence of single TLRs or the TLR signalling adaptor protein MyD88, suggesting a normal costimulation of T cells (Nicolle et al 2004a, Fremond et al 2004).

In summary, these results suggest that purified mycobacterial antigens and whole bacilli preferentially interact with TLR2 and TLR4, possibly in combination with additional TLRs and PRRs, leading to MyD88-dependent activation of anti-bacterial effector pathways.

Role of TLR *in vivo* infection with *Mycobacterium bovis* BCG

To test the role of TLR signalling *in vivo* in controlling mycobacterial infection we have performed studies in TLR gene deficient mice.

Infection with *Mycobacterium bovis* (BCG) caused persistent inflammation in C3H/HeJ TLR4 mutant mice, and therefore TLR4 is not required to control acute BCG infection, but may have a function for the fine tuning of inflammation in chronic infection (Fremond et al 2003). At high infectious dose bacterial clearance and reduced IFNγ secretion was reported in TLR deficient mice (Heldwein et al 2003). Further, BCG infection resolved in the chronic phase in TLR2-deficient mice (Nicolle et al 2004b). Interestingly, the adaptive response of TLR2- and/or TLR4-deficient mice seemed essentially normal on day 14 or 56 after infection, as T cells responded normally to soluble BCG antigens unlike previously reported (Heldwein et al 2003). In conclusion, our data demonstrate that TLR2, TLR4 or TLR6 are redundant for the control of *M. bovis* BCG mycobacterial infection.

To assess the role of a global TLR signalling in host response to mycobacterial infection, we infected mice deficient in the TLR adaptor molecule myeloid differentiation factor 88 (MyD88) with the vaccine strain *M. bovis* BCG, and the immune response and bacterial burden were investigated. BCG (2×10^6 CFU i.v.) infected MyD88-deficient mice had increased lung weights at 8 months with confluent chronic pneumonia and two log higher CFU in the lung than wild-type mice (Nicolle et al 2004a), while the infection was controlled in liver and spleen and there was efficient systemic T cell priming with high IFNγ production by $CD4^+$ splenic T cells in MyD88-deficient mice. Lung infiltrating cells showed IFNγ production by pulmonary $CD4^+$ T cells upon specific restimulation, and a reduced capacity to produce nitric oxide and IL10. In summary, despite the dramatic reduction of the innate immune response, MyD88-deficient mice were able to mount an efficient T cell response to mycobacterial antigens, which was however insufficient to control infection in the lung, resulting in chronic pneumonia in MyD88 deficient mice (Nicolle et al 2004b).

These results are surprising as BCG was cleared from all organs, but was capable of surviving in the lung in the absence of TLR–MyD88 signalling. However, as the absence of single TLRs had only a minor effect *in vivo* in response to the attenuated *M. bovis* BCG vaccine strain, we tested a virulent *Mtb* strain.

Critical role of TLR signalling in *Mtb* infection

Using single TLR-deficient mice we investigated their susceptibility to low dose aerosol infection with virulent *Mtb* H37Rv. TLR4 deficient mice displayed reduced bacterial clearance during a long-term infection protocol and developed a chronic pneumonia and died within 15 weeks (Abel et al 2002). The data were confirmed recently by an independent group (Branger et al 2004). In short-term infections, no significant differences in the inflammatory response or the bacterial burden in infected organs during the first 50 days of infection and long-term were found (Reiling et al 2002, Shim et al 2003, Kamath et al 2003). CD14, a coreceptor of TLR4, appears not to be involved in host resistance, as CD14-deficient mice clear the infection normally (Reiling et al 2002) (M. Jacobs, unpublished data).

Then we investigated the role of TLR2 in the host response and infected TLR2-deficient mice by aerosol using 500 CFU. TLR2-deficient mice initially control an aerosol infection with signs of T cell activation, but develop increased bacterial burden and chronic pneumonia with death in 5 months (Drennan et al 2004). Although inflammatory cells such as macrophages and activated T cells are recruited, no distinct granulomas are formed in TLR2-deficient mice. Inflammation in the presence of a high bacterial load is associated with increased TNFα, IL12p40 and IFNγ production in the lung (Drennan et al 2004). Therefore, the data suggest that TLR2 may function as a regulator of inflammation, and in its absence an exaggerated immune-inflammatory response develops. By contrast, others found only a minor role for TLR2 in the control of *Mtb* infection (Reiling et al 2002, Sugawara et al 2003).

Since TLR2 forms heterodimers with either TLR6 or TLR1, we asked whether the co-receptors are involved in mycobacterial responses. TLR6-deficient mice are resistant to high *Mtb* aerosol infection (Gomes et al 1999), and no data are so far available on the role of TLR1 signalling in the *in vivo* host response. Since at least TLR2 and TLR4 are modulating host resistance to mycobacterial infection, we asked whether inactivation of both TLR2 and TLR4 augment the susceptibility to *Mtb* infection, which is indeed the case (Fremond et al unpublished results 2006).

Finally, we investigated the role of the common TLR adaptor molecule MyD88. We infected MyD88-deficient mice with *Mtb* (Fremond et al 2004). Aerogenic infection of MyD88-deficient mice with *Mtb* was lethal within 4 weeks as shown for TNF-deficient mice. Mice succumbed to acute necrotic pneumonia with 2 log higher CFU in the lung (Fig. 2). This was associated with high pulmonary levels of cytokines and chemokines, and acute, necrotic pneumonia, despite a normal T cell response with IFNγ production to mycobacterial antigens upon *ex vivo* restimulation. The phenotype resembles that of TNF deficiency (Flynn et al 1995, Jacobs et al 2000a). Fatal infection in MyD88-deficient mice is not surprising in view of their profound defect of innate immunity (Shi et al 2003, Scanga et al 2004).

We then asked whether MyD88-deficient mice are able to develop an adaptive immune response. In view of the fact that MyD88-deficient macrophages up-regulated costimulatory molecules such as CD40 and CD86 (Nicolle et al 2004a), we investigated the response to BCG vaccination. We found comparable antigen

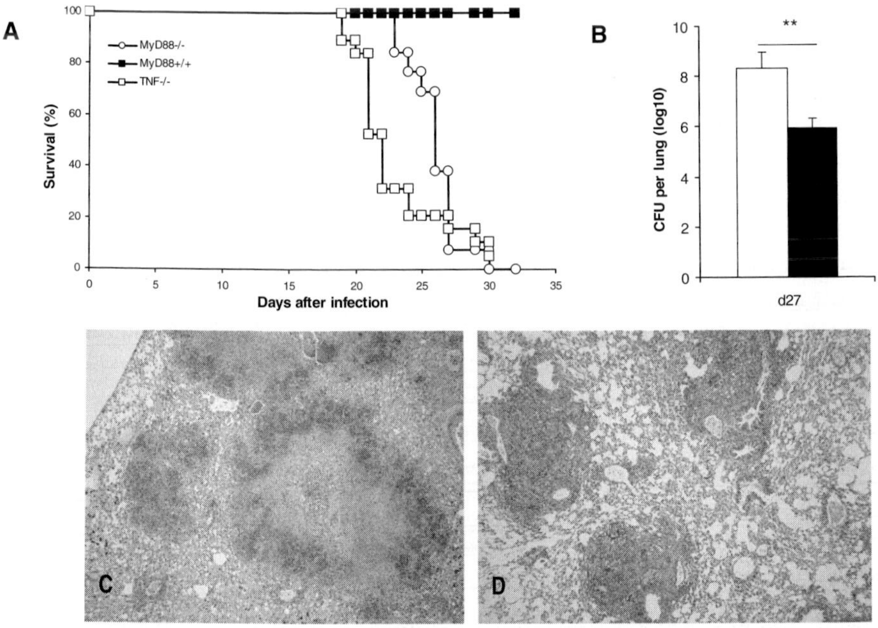

FIG. 2. MyD88-deficient mice are unable to clear virulent *M. tuberculosis* H37Rv and succumb to infection with acute necrotic pneumonia. Survival of MyD88- and TNF-deficient mice as compared with controls. (A) Mycobacterial burden (CFO) in the lung (B) and lung pathology: Acute necrotic pneumonia MyD88 deficient mice (C) and typical granulomas in wild-type mice (D). Wild-type, MyD88- and TNF-deficient mice were infected with *M. tuberculosis* H37Rv (200 CFU/n) and sacrificed at 27 days (modified from Fremond et al 2004).

specific responses with IFNγ production in the absence of MyD88 as in wild-type mice.

Furthermore, prior BCG vaccination conferred a substantial protection in vaccinated MyD88$^{-/-}$ mice from acute *Mtb* infection with 2 log reduced mycobacterial load (CFU) in the lung as compared to non-vaccinated mice (Fremond et al 2004). These data demonstrate that MyD88 signalling is dispensable to raise an acquired immune response, which however is not sufficient to compensate the profound innate immune defect of MyD88-deficient mice to control *Mtb* infection.

In conclusion, the present *in vivo* evidence suggests that signalling through single TLRs has only a modest effect in acute mycobacterial infection, while abrogation of most of TLR signalling as found in MyD88-deficient mice results in profound deficiency of innate immunity with preserved adaptive immunity.

Role of other pattern recognition receptors in the control of mycobacterial infection

Recognition of pathogens is however not limited to TLRs. TLR-unrelated receptors such as the CD14, scavenger and complement receptors, pulmonary surfactant protein A, DC-SIGN (DC-specific intercellular adhesion molecule-3 grabbing nonintegrin), CD40 and CD44 have been implicated in recognition of mycobacterial antigens and coupling a cellular response. The evidence will be discussed briefly.

Nucleotide-binding oligomerization domain proteins (NOD) belong to a TLR related protein family with leucine-rich repeats which likely have a role in the intracellular recognition of pathogen ligands such as peptidoclycans, muramyl dipeptides and diaminopimelate-containing *N*-acetylglucosamine-*N*-acetylmuramic acid (GlcNAc-MurNAc) tripeptide (Inohara & Nunez 2003, Chamaillard et al 2003, Girardin et al 2003a). NOD2 has been linked to the inflammatory bowel disorder Crohn's disease (Girardin et al 2003b). *Mycobacterium avium* subspecies paratuberculosis (MAP) is presently the most favourite pathogen linked to Crohn's disease (Greenstein 2003). Except for the circumstantial evidence for MAP, the role of NOD proteins in response to mycobacterial antigen is so far unknown.

Conclusion

The present results support the emerging paradigm that at least TLR2 and TLR4 play a role in sensing mycobacteria and mounting an antimycobacterial immune response *in vitro* and *in vivo*. In future, the repertoire of identified mycobacterial antigens which mediate TLR signalling is likely to continue to increase. The combinatorial recognition of pathogen-associated molecular patterns by more than one TLR, and cross-talk between TLRs and other PRRs opens a new dimension

(Underhill 2003). A more detailed knowledge of the stimulatory and inhibitory ligands of TLRs and PRRs might allow (Janeway & Medzhitov 2002) modulation of the immune response to mycobacteria. In summary, the common IL1R TLR adaptor molecule MyD88 is critical to develop a robust host response to *Mtb* infection, and the profound defect of innate immunity is not compensated by other molecules. Interestingly, MyD88 deficiency allows the emergence of an adaptive immunity. Therefore IL1R TLRs contribute to an efficient innate response to mycobacteria, while MyD88-dependent signalling is dispensable for adaptive immunity.

References

Abel B, Thieblemont N, Quesniaux VJ et al 2002 Toll-like receptor 4 expression is required to control chronic mycobacterium tuberculosis infection in mice. J Immunol 169:3155–3162

Akira S, Takeda K 2004 Toll-like receptor signalling. Nat Rev Immunol 4:499–511

Akira S, Yamamoto M, Takeda K 2003 Role of adapters in toll-like receptor signalling. Biochem Soc Trans 31:637–642

Aliprantis AO, Yang RB, Mark MR et al 1999 Cell activation and apoptosis by bacterial lipoproteins through toll-like receptor-2. Science 285:736–739

Barnes PF, Chatterjee D, Abrams JS et al 1992 Cytokine production induced by mycobacterium tuberculosis lipoarabinomannan. Relationship to chemical structure. J Immunol 149:541–547

Botha T, Ryffel B 2003 Reactivation of latent tuberculosis infection in TNF-deficient mice. J Immunol 171:3110–3118

Branger J, Leemans JC, Florquin S et al 2004 Toll-like receptor 4 plays a protective role in pulmonary tuberculosis in mice. Int Immunol 16:509–516

Brightbill HD, Libraty DH, Krutzik SR et al 1999 Host defense mechanisms triggered by microbial lipoproteins through toll-like receptors. Science 285:732–736

Bulut Y, Faure E, Thomas L, Equils O, Arditi M 2001 Cooperation of Toll-like receptor 2 and 6 for cellular activation by soluble tuberculosis factor and Borrelia burgdorferi outer surface protein A lipoprotein: role of Toll-interacting protein and IL-1 receptor signaling molecules in Toll-like receptor 2 signaling. J Immunol 167:987–994

Chamaillard M, Hashimoto M, Horie Y et al 2003 An essential role for NOD1 in host recognition of bacterial peptidoglycan containing diaminopimelic acid. Nat Immunol 4:702–707

Chan J, Tanaka K, Carroll D, Flynn J, Bloom BR 1995 Effects of nitric oxide synthase inhibitors on murine infection with mycobacterium tuberculosis. Infect Immun 63:736–740

Chatterjee D, Khoo KH 1998 Mycobacterial lipoarabinomannan: an extraordinary lipoheteroglycan with profound physiological effects. Glycobiology 8:113–120

Cooper AM, Kipnis A, Turner J et al 2002 Mice lacking bioactive IL-12 can generate protective, antigen-specific cellular responses to mycobacterial infection only if the IL-12 p40 subunit is present. J Immunol 168:1322–1327

Daffe M, Draper P 1998 The envelope layers of mycobacteria with reference to their pathogenicity. Adv Microb Physiol 39:131–203

Drennan MB, Nicolle D, Quesniaux VJ et al 2004 Toll-like receptor 2-deficient mice succumb to mycobacterium tuberculosis infection. Am J Pathol 164:49–57

Ehlers S, Holscher C, Scheu S et al 2003 The lymphotoxin beta receptor is critically involved in controlling infections with the intracellular pathogens Mycobacterium tuberculosis and listeria monocytogenes. J Immunol 170:5210–5218

Ehrt S, Schnappinger D, Bekiranov S et al 2001 Reprogramming of the macrophage transcriptome in response to interferon-gamma and Mycobacterium tuberculosis: signaling roles of nitric oxide synthase-2 and phagocyte oxidase. J Exp Med 194:1123–1140

Flynn JL 2004 Immunology of tuberculosis and implications in vaccine development. Tuberculosis (Edinb) 84:93–101

Flynn JL, Chan J 2001a Immunology of tuberculosis. Annu Rev Immunol 19:93–129

Flynn JL, Chan J 2001b Tuberculosis: latency and reactivation. Infect Immun 69:4195–4201

Flynn JL, Chan J 2003 Immune evasion by Mycobacterium tuberculosis: living with the enemy. Curr Opin Immunol 15:450–455

Flynn JL, Goldstein MM, Chan J et al 1995 Tumor necrosis factor-alpha is required in the protective immune response against Mycobacterium tuberculosis in mice. Immunity 2:561–572

Flynn JL, Scanga CA, Tanaka KE, Chan J 1998 Effects of aminoguanidine on latent murine tuberculosis. J Immunol 160:1796–803

Fremond CM, Nicolle DM, Torres DS, Quesniaux VF 2003 Control of Mycobacterium bovis BCG infection with increased inflammation in TLR4-deficient mice. Microbes Infect 5:1070–1081

Fremond C, Yeremeev V, Nicolle D et al 2004 Fatal Mycobacterium tuberculosis infection despite adaptive immune response in the absence of MyD88. J Clin Invest 114:1790–1799

Garcia I, Guler R, Vesin D et al 2000 Lethal Mycobacterium bovis Bacillus Calmette Guerin infection in nitric oxide synthase 2-deficient mice: cell-mediated immunity requires nitric oxide synthase 2. Lab Invest 80:1385–1397

Gehring AJ, Rojas RE, Canaday DH et al 2003 The Mycobacterium tuberculosis 19-kilodalton lipoprotein inhibits gamma interferon-regulated HLA-DR and Fc gamma R1 on human macrophages through Toll-like receptor 2. Infect Immun 71:4487–4497

Gehring AJ, Dobos KM, Belisle JT, Harding CV, Boom WH 2004 Mycobacterium tuberculosis LprG (Rv1411c): a novel TLR-2 ligand that inhibits human macrophage class II MHC antigen processing. J Immunol 173:2660–2668

Gilleron M, Quesniaux VF, Puzo G 2003 Acylation state of the phosphatidylinositol hexamannosides from Mycobacterium bovis bacillus Calmette Guerin and mycobacterium tuberculosis H37Rv and its implication in Toll-like receptor response. J Biol Chem 278: 29880–29889

Girardin SE, Boneca IG, Carneiro LA et al 2003a Nod1 detects a unique muropeptide from gram-negative bacterial peptidoglycan. Science 300:1584–1587

Girardin SE, Hugot JP, Sansonetti PJ 2003b Lessons from Nod2 studies: towards a link between Crohn's disease and bacterial sensing. Trends Immunol 24:652–658

Gomes MS, Florido M, Pais TF, Appelberg R 1999 Improved clearance of Mycobacterium avium upon disruption of the inducible nitric oxide synthase gene. J Immunol 162:6734–6739

Gomez JE, McKinney JD 2004 M. tuberculosis persistence, latency, and drug tolerance. Tuberculosis (Edinb) 84:29–44

Greenstein RJ 2003 Is Crohn's disease caused by a mycobacterium? Comparisons with leprosy, tuberculosis, and Johne's disease. Lancet Infect Dis 3:507–514

Hajjar AM, O'Mahony DS, Ozinsky A et al 2001 Cutting edge: functional interactions between toll-like receptor (TLR) 2 and TLR1 or TLR6 in response to phenol-soluble modulin. J Immunol 166:15–19

Heldwein KA, Liang MD, Andresen TK et al 2003 TLR2 and TLR4 serve distinct roles in the host immune response against Mycobacterium bovis BCG. J Leukoc Biol 74:277–286

Hoebe K, Du X, Georgel P et al 2003 Identification of Lps2 as a key transducer of MyD88-independent TIR signalling. Nature 424:743–748

Holscher C, Atkinson RA, Arendse B et al 2001 A protective and agonistic function of IL-12p40 in Mycobacterial infection. J Immunol 167:6957–6966

Horng T, Barton GM, Flavell RA, Medzhitov R 2002 The adaptor molecule TIRAP provides signalling specificity for Toll-like receptors. Nature 420:329–333

Inohara N, Nunez G 2003 NODs: intracellular proteins involved in inflammation and apoptosis. Nat Rev Immunol 3:371–382

Jacobs M, Brown N, Allie N, Ryffel B 2000a Fatal Mycobacterium bovis BCG infection in TNF-LT-alpha-deficient mice. Clin Immunol 94:192–199

Jacobs M, Marino MW, Brown N et al 2000b Correction of defective host response to Mycobacterium bovis BCG infection in TNF-deficient mice by bone marrow transplantation. Lab Invest 80:901–914

Janeway CA Jr, Medzhitov R 2002 Innate immune recognition. Annu Rev Immunol 20:197–216

Jones BW, Heldwein KA, Means TK, Saukkonen JJ, Fenton MJ 2001a Differential roles of Toll-like receptors in the elicitation of proinflammatory responses by macrophages. Ann Rheum Dis 60 Suppl 3:iii6–12

Jones BW, Means TK, Heldwein KA et al 2001b Different Toll-like receptor agonists induce distinct macrophage responses. J Leukoc Biol 69:1036–1044

Kamath AB, Alt J, Debbabi H, Behar SM 2003 Toll-like receptor 4-defective C3H/HeJ mice are not more susceptible than other C3H substrains to infection with Mycobacterium tuberculosis. Infect Immun 71:4112–4118

Lazarevic V, Myers AJ, Scanga CA, Flynn JL 2003 CD40, but not CD40L, is required for the optimal priming of T cells and control of aerosol M. tuberculosis infection. Immunity 19:823–835

Means TK, Lien E, Yoshimura A et al 1999a The CD14 ligands lipoarabinomannan and lipopolysaccharide differ in their requirement for Toll-like receptors. J Immunol 163:6748–6755

Means TK, Wang S, Lien E et al 1999b Human toll-like receptors mediate cellular activation by Mycobacterium tuberculosis. J Immunol 163:3920–3927

Means TK, Jones BW, Schromm AB et al 2001 Differential effects of a Toll-like receptor antagonist on Mycobacterium tuberculosis-induced macrophage responses. J Immunol 166:4074–4082

Medzhitov R, Preston-Hurlburt P, Janeway CA Jr 1997 A human homologue of the Drosophila Toll protein signals activation of adaptive immunity. Nature 388:394–397

Mohan VP, Scanga CA, Yu K et al 2001 Effects of tumor necrosis factor alpha on host immune response in chronic persistent tuberculosis: possible role for limiting pathology. Infect Immun 69:1847–1855

Nicolle DM, Pichon X, Bouchot A et al 2004a Chronic pneumonia despite adaptive immune response to Mycobacterium bovis BCG in MyD88-deficient mice. Lab Invest 84:1305–1321

Nicolle DM, Fremond C, Pichon X et al 2004b Long-term control of Mycobacterium bovis BCG infection in the absence of Toll-Like receptors: investigation on TLR2, TLR6 or TLR2–TLR4 deficient mice. Infect Immun 72:6994–7004

Nigou J, Gilleron M, Rojas M et al 2002 Mycobacterial lipoarabinomannans: modulators of dendritic cell function and the apoptotic response. Microbes Infect 4:945–953

North RJ, Jung YJ 2004 Immunity to tuberculosis. Annu Rev Immunol 22:599–623

Noss EH, Harding CV, Boom WH 2000 Mycobacterium tuberculosis inhibits MHC class II antigen processing in murine bone marrow macrophages. Cell Immunol 201:63–74

Noss EH, Pai RK, Sellati TJ et al 2001 Toll-like receptor 2-dependent inhibition of macrophage class II MHC expression and antigen processing by 19-kDa lipoprotein of Mycobacterium tuberculosis. J Immunol 167:910–918

O'Neill LA, Fitzgerald KA, Bowie AG 2003 The Toll-IL-1 receptor adaptor family grows to five members. Trends Immunol 24:286–289

Oshiumi H, Matsumoto M, Funami K, Akazawa T, Seya T 2003 TICAM-1, an adaptor molecule that participates in Toll-like receptor 3-mediated interferon-beta induction. Nat Immunol 4:161–167

Pai RK, Convery M, Hamilton TA, Boom WH, Harding CV 2003 Inhibition of IFN-gamma-induced class II transactivator expression by a 19-kDa lipoprotein from Mycobacterium tuberculosis: a potential mechanism for immune evasion. J Immunol 171:175–184

Quesniaux V, Ryffel B 2004 Toll-like receptors: emerging targets of immunomodulation. Expert Opin Ther Patents 14:85–100

Quesniaux V, Fremond C, Jacobs M et al 2004a Toll-like receptor pathways in the immune responses to mycobacteria. Microbes Infect 6:946–959

Quesniaux VJ, Nicolle DM, Torres D et al 2004b Toll-like receptor 2 (TLR2)-dependent-positive and TLR2-independent-negative regulation of proinflammatory cytokines by mycobacterial lipomannans. J Immunol 172:4425–4434

Reed MB, Domenech P, Manca C et al 2004 A glycolipid of hypervirulent tuberculosis strains that inhibits the innate immune response. Nature 431:84–87

Reiling N, Holscher C, Fehrenbach A et al 2002 Cutting edge: Toll-like receptor (TLR)2- and TLR4-mediated pathogen recognition in resistance to airborne infection with Mycobacterium tuberculosis. J Immunol 169:3480–3484

Roach DR, Briscoe H, Baumgart K, Rathjen DA, Britton WJ 1999 Tumor necrosis factor (TNF) and a TNF-mimetic peptide modulate the granulomatous response to Mycobacterium bovis BCG infection in vivo. Infect Immun 67:5473–5476

Roach DR, Briscoe H, Saunders B et al 2001 Secreted lymphotoxin-alpha is essential for the control of an intracellular bacterial infection. J Exp Med 193:239–246

Ryffel B, Fremond C, Jacobs M et al 2005 Innate immunity to mycobacterial infection in mice: critical role for Toll-like receptors. Tuberculosis (Edinb) 85:395–405

Scanga CA, Mohan VP, Joseph H et al 1999 Reactivation of latent tuberculosis: variations on the Cornell murine model. Infect Immun 67:4531–4538

Scanga CA, Mohan VP, Yu K et al 2000 Depletion of CD4(+) T cells causes reactivation of murine persistent tuberculosis despite continued expression of interferon gamma and nitric oxide synthase 2. J Exp Med 192:347–358

Scanga CA, Bafica A, Feng CG et al 2004 MyD88-deficient mice display a profound loss in resistance to mycobacterium tuberculosis associated with partially impaired Th1 cytokine and nitric oxide synthase 2 expression. Infect Immun 72:2400–2404

Shi S, Nathan C, Schnappinger D et al 2003 MyD88 primes macrophages for full-scale activation by interferon-γ yet mediates few responses to Mycobacterium tuberculosis. J Exp Med 198:987–997

Shim TS, Turner OC, Orme IM 2003 Toll-like receptor 4 plays no role in susceptibility of mice to Mycobacterium tuberculosis infection. Tuberculosis (Edinb) 83:367–371

Sugawara I, Yamada H, Li C et al 2003 Mycobacterial infection in TLR2 and TLR6 knockout mice. Microbiol Immunol 47:327–336

Takeda K, Akira S 2004 TLR signaling pathways. Semin Immunol 16:3–9

Underhill DM 2003 Mini-review Toll-like receptors: networking for success. Eur J Immunol 33:1767–1775

Underhill DM, Bassetti M, Rudensky A, Aderem A 1999 Dynamic interactions of macrophages with T cells during antigen presentation. J Exp Med 190:1909–1914

Vignal C, Guerardel Y, Kremer L et al 2003 Lipomannans, but not lipoarabinomannans, purified from Mycobacterium chelonae and Mycobacterium kansasii induce TNF-alpha and IL-8 secretion by a CD14-toll-like receptor 2-dependent mechanism. J Immunol 171:2014–2023

Yamamoto M, Sato S, Hemmi H et al 2002a Essential role for TIRAP in activation of the signalling cascade shared by TLR2 and TLR4. Nature 420:324–329

Yamamoto M, Sato S, Mori K et al 2002b Cutting edge: a novel Toll/IL-1 receptor domain-containing adapter that preferentially activates the IFN-beta promoter in the Toll-like receptor signaling. J Immunol 169:6668–6672

Yamamoto M, Sato S, Hemmi H et al 2003 Role of adaptor TRIF in the MyD88-independent Toll-like receptor signaling pathway. Science 301:640–643

Zhang D, Zhang G, Hayden MS et al 2004 A toll-like receptor that prevents infection by uropathogenic bacteria. Science 303:1522–1126

DISCUSSION

Speert: A similar thing has been found with *Pseudomonas* in cystic fibrosis. The state of acylation of LPS influences the extent of inflammation in human cells but not in murine cells. Does this hold for both mouse and human? Second, if you follow the organism during the course of infection from acute to chronic phase, does the degree of acylation change?

Quesniaux: It's difficult to say in the course of the infection, to ask the biochemists to purify the LM/LAM forms from murine tissues in well controlled mouse infections. We hope to do this in bovine models.

Speert: Were your data purely from murine studies?

Quesinaux: Yes.

Latgé: This reminds me of the antigenic variation in parasites.

Wilkinson: I am interested to see you describe the pneumonia that occurs in both MyD88- and TNF-deficient mice as necrotic. Is it truly necrotic, or an alveolitis?

Ryffel: It is clearly an acute pneumonia with confluent necrosis reminding of caseation, which we find in MyD88, TNF and lymphotoxin knockout mice which develop uncontrolled infection, with extensive and necrotic pneumonia (Jacobs et al 2002, Fremond et al 2004, 2005).

Wilkinson: Necrosis in TB has been ascribed classically to TNF. So what drives the necrosis in TNF-deficient mice?

Ryffel: That's a difficult question. If we look at macrophage *in vitro* there is no proinflammatory response. TNF and other cytokines have alternative pathways. At 4 weeks of infection cytokine responses are correlated with bacterial load. It is possible that there is a combination of interferon, TNF and IL12 which may cause necrosis. I don't think we can pinpoint TNF only in this process.

Gordon: Is there apoptosis in the mouse model?

Ryffel: Yes, along with necrosis there is apoptosis as shown by caspase staining.

Steinman: What do you know about the resistance of the MyD88 knockout mice that are vaccinated with BCG? Do you know that this is $CD4^+$ T cell dependent? Have you looked if innate lymphocytes are responsible?

Ryffel: That is an important question. We haven't done any transfer experiments, but we should do. The fact that we find $CD4^+$ cells expressing IFNγ *ex vivo* is not sufficient.

Steinman: People are looking for a role for the CD1-dependent presentation of glycolipids. Perhaps this will be seen in a MyD88-deficient background.

Gordon: What do those people more interested in the human disease think about this? Is this relevant to anything? Has anyone looked at Toll-like receptors (TLR) polymorphisms in susceptibility, and what is the outcome?

Wilkinson: There are some data but the associations are moderate.

Quesniaux: There are some published data on TLR2 polymorphisms: R753Q is associated with higher predisposition for *Mtb* and Staphylococci infections while R667W seems associated with increased *Mtb* and *M. leprae* infections (Cook et al 2004).

Segal: There is a linkage of TLR4 to Crohn's disease, particularly of the large bowel (Franchimont et al 2004). I think it is a secondary involvement.

Gordon: Ultimately, there will be quite a lot of these claims. What does one make of them?

Speert: The issue of looking at human cells as regards their response to the degree of acylation of *Pseudomonas aeruginosa* is interesting, as inflammation is dramatically different between human and murine cells. During chronic infection in cystic fibrosis the degree of acylation increases, and there is no change on the effects on the mouse, but the effect on human cells is dramatic.

Quesniaux: The proinflammatory effect has also been looked at in a human monocytic THP1 cell line.

Didierlaurent: In the MyD88 knockout mouse there are some defects in cell proliferation in the gut. Have you looked at the basic function of the lungs in these mice?

Ryffel: I don't see a defect, but any proinflammatory response is dramatically reduced. In the gut a critical role of MyD88 in controlling commensal bacteria entry has been demonstrated (Rakoff-Nahoum et al 2004). The lung function in the absence of MyD88 appears normal, and the response to inhaled endotoxin is completely ablated (Noulin et al 2005).

Gordon: So the idea is coming out that the TLRs are necessary for integrity of epithelia in the gut and perhaps the lung also. There's a recent paper in *Nature Medicine* on this (Jiang et al 2005). I don't know how you assay for this easily.

Didierlaurent: The infection is long-term so the absence of TLR signalling may have some effect on lung repair mechanisms occurring after the initial inflammatory response.

Ryffel: The way we plan to address this is by using Cre-Lox for MyD88 to inactivate it in the lung.

References

Cook DN, Pisetsky DS, Schwartz DA 2004 Toll-like receptors in the pathogenesis of human disease. Nat Immunol 5:975–979

Franchimont D, Vermeire S, El Housni H et al 2004 Deficient host-bacteria interactions in inflammatory bowel disease? The toll-like receptor (TLR)-4 Asp299gly polymorphism is associated with Crohn's disease and ulcerative colitis. Gut 53:987–992

Fremond CM, Yeremeev V, Nicolle DM, Jacobs M, Quesniaux VF, Ryffel B 2004 Fatal Mycobacterium tuberculosis infection despite adaptive immune response in the absence of MyD88. J Clin Invest 114:1790–1799

Fremond C, Allie N, Dambuza I et al 2005 Membrane TNF confers protection to acute mycobacterial infection. Respir Res 6:136

Jacobs M, Fick L, Allie N, Brown N, Ryffel B 2002 Enhanced immune response in Mycobacterium bovis bacille Calmette Guerin (BCG)-infected IL-10-deficient mice. Clin Chem Lab Med 40:893–902

Jiang D, Liang J, Fan J et al 2005 Regulation of lung injury and repair by Toll-like receptors and hyaluronan. Nat Med 11:1173–1179

Noulin N, Quesniaux VF, Schnyder-Candrian S et al 2005 Both hemopoietic and resident cells are required for MyD88-dependent pulmonary inflammatory response to inhaled endotoxin. J Immunol 175:6861–6869

Rakoff-Nahoum S, Paglino J, Eslami-Varzaneh F, Edberg S, Medzhitov R 2004 Recognition of commensal microflora by toll-like receptors is required for intestinal homeostasis. Cell 118:229–241

Population of lungs by mast cells

T. J. Williams and C. L. Weller

Leukocyte Biology Section, National Heart and Lung Institute, Faculty of Medicine, Sir Alexander Fleming Building, Imperial College London, South Kensington Campus, London SW7 2AZ, UK

Abstract. Mast cells are tissue-resident cells that are localized particularly in the skin, gastrointestinal tract and respiratory tract. They are mainly recognised for their role in adaptive immunity and allergy where cross-linking of surface-bound IgE results in acute mediator release giving early symptoms, and cytokine production contributing to chronic changes. The mast cell is now also increasingly recognized for its role in innate immunity conferred by its repertoire of complement and Toll receptors. Thus, mast cell deletion has been shown to suppress certain innate immune responses in murine models. Our interest is in the mechanisms involved in population of tissues by mast cells, particularly the airways. Mast cells are released from the bone marrow into the blood as committed precursors. These cells circulate in very low numbers and accumulate in tissues where they proliferate and mature under the influence of local cytokines and growth factors that define the mature phenotype appropriate for their location. Chemoattraction is important at critical phases in the life history of the mast cell, i.e. movement towards and through the bone marrow sinus endothelium, recruitment to tissues and movement within the tissues to the location of the mature cell. These phases are dependent on chemoattractants generated at specific locations acting on cell surface receptors whose repertoire evolves as the mast cell matures.

2006 Innate immunity to pulmonary infection. Wiley, Chichester (Novartis Foundation Symposium 279) p 142–154

Mast cells in innate immunity

Mast cells are long-lived tissue-resident cells that are prominent at the interface between the body and its environment, the skin, gastrointestinal tract and respiratory tract. They are most studied because of their important role in adaptive immunity to helminth infection and in allergic reactions. In this context, specificity is conferred by IgE antibody bound to high affinity FcεRI receptors on the mast cell surface. Cross-linking by specific antigens triggers the release of preformed mediators (such as histamine and proteases), *de novo* synthesis of acute mediators such as leukotrienes, and synthesis of cytokines. The acute mediators induce responses in endothelial cells, smooth muscle cells and glands that result in the acute symptoms of infection or allergy. However, the mast cell is increasingly recognized for its role in more chronic symptoms, where cytokine production is important.

Although mast cells are identified more readily with these adaptive responses, research, particularly over the past 10 years, has shown that mast cells are also important in innate immunity. A long-established link is via the complement system. The complement fragments, C3a and C5a, are induced as a consequence of antigen-specific responses via the classical activation pathway, and during alternative pathway activation induced by components of microbial cell walls, such as lipopolysaccharide (LPS), as part of innate immunity. Mast cells have C3a and C5a receptors that trigger cell activation following ligand binding to initiate inflammatory responses during injury or infection. In addition, mast cells have been shown to have a novel collectin/C1q receptor that is able to trigger mast cell activation and innate immunity (Edelson et al 2005).

Mast cells are also able to respond directly to LPS that was shown in early studies to stimulate cytokine production, without acute histamine release (Leal-Burumen et al 1994). Mast cells have been reported to express Toll-like receptors (TLRs) 1, 2, 3, 4, 6 and 9 (Applequist et al 2002, Marshall & Jawdat 2004) with the precise repertoire depending on species and type of mast cell. Definitive evidence of an important role for mast cells in innate immunity came from experiments *in vivo*. Mice deficient in mast cells were shown to die from peritonitis induced by caecal ligation and puncture (Malaviya et al 1996) or *Klebsiella pneumoniae* infection (Echtenacher et al 1996) whereas normal mice survived. Survival could be restored in deficient mice by administering mast cells into the peritoneal cavity. Several pathways of mast cell stimulation via TLRs have been established from experiments *in vitro*. One example is the stimulation of mast cells by LPS from *Escherichia coli* via TLR4 that results in production of tumour necrosis factor (TNF)α, interleukin (IL)1β, IL6 and IL13, but not IL4 or IL5 (Supajatura et al 2002). In contrast, peptidoglycan from *Staphylococcus aureus* stimulates mast cells via TLR2 resulting in TNFα, IL4, IL5, IL6 and IL13, but not IL1β. Further, stimulation via TLR2, but not TLR4, results in calcium mobilization and mast cell degranulation.

In addition to their role in bacterial infection, mast cells are also recognized for their role in host defence to viruses. For example, HIV and dengue virus can infect mast cells and induce degranulation, and the production of cytokines and chemokines (Marshall & Jawdat 2004). Mast cells have also been implicated in host responses to respiratory syncytial virus and Sendae virus.

The localization of mast cells in a first-line defence position gives them the potential to respond early to infectious agents and then to orchestrate the recruitment and activation of other cells to continue the defence process, thus acting as sentinel cells. This involves several stages. The first is recognition as outlined above, either directly involving TLRs, or indirectly via complement receptors. The next phase is the mast cell response. Mast cells are capable of phagocytosis (Arock et al 1998) and they can also secrete antimicrobial peptides such as cathelicidin-related antimicrobial peptide or β-defensins (Di Nardo et al 2003). The acute mediators produced on mast cell activation also stimulate responses in other tissue cells that can influence early phases of host defence, e.g. histamine and leukotrienes acting on the microvascular endothelium to induce plasma protein leakage (thus increasing the supply of complement components to tissues) and increased intestinal motility. As mentioned above, mast cells are able to produce a different selection of mediators depending on the nature of the stimulus. In several murine infection models mast cell-derived TNFα seems to be particularly important in mediating the recruitment of neutrophils, thus mobilizing cells with phagocytic and microbial killing activity. Mast cells can also produce chemokines that can recruit other cell types, such as CCL20, that can induce the accumulation of immature dendritic cells and T lymphocytes (Lin et al 2003).

Thus, the mast cell has an important function as a sentinel cell. Its defensive role is dependent on its location and hence ultimately on chemoattractant mechanisms. This is the subject of current investigations in our laboratory. Most of the research in this area is on mast cell recruitment in allergic reactions or in models of helminth infection. However, these mechanisms are also relevant to innate immunity.

The origin of mast cells

Mast cells are derived from pluripotential stem cells in the bone marrow. Under the influence of specific growth factors and cytokines, these cells mature into inflammatory cells specialised to fulfil defined roles in host defence and tissue repair. Mast cells are released from the bone marrow into the blood as progenitors, which are recruited into tissues, where they mature into long-lived resident tissue cells. Reconstitution studies in mast cell-deficient mice have produced considerable information about the origin of mast cells (Kitamura et al 1978). Mouse bone marrow cultures supplemented with IL3 result in an enriched mast cell population, with the cells characterised as FcεRI$^+$ and c-kit^{high} (tyrosine kinase receptor for stem cell factor, SCF) reaching 85% purity at 3 weeks, and these cells are able to populate peripheral tissues on reconstitution. The numbers of mast cell progenitors have been estimated at approximately 10–70 per 10^5 bone marrow cells and at 1–2 per 10^5 nucleated cells in blood (Sonoda et al 1982). Studies of mast cell progenitors in tissues are difficult because of their very low numbers *in situ*. A minor population of circulating c-kit$^+$ committed mast cell progenitors has been reported in mouse fetal blood (Rodewald et al 1996). More recently, sequential immunomagnetic isolation of adult mouse bone marrow has revealed a 0.02% population of undifferentiated mast cells characterised as CD34$^+$, CD13$^+$, c-kit$^+$ and FcεR1$^-$ (Jamur et al 2005). A second study identified mast cell progenitors in adult mouse bone marrow, as Lin$^-$, c-kit$^+$, Sca-1$^-$, Ly6c$^-$, FcεR1α^-, CD27$^-$, $\beta 7^+$ and T1/ST2$^+$ (Chen et al 2005). Other studies using limiting dilution assays have been employed to determine the numbers of mast cell progenitors in different tissues, including the intestines, lung, spleen and bone marrow (Gurish et al 2001). A c-kit^{+} α_4^{hi} β_7^{+} mast cell progenitor has been reported in mouse bone marrow 5 days after infection of the small intestine with *Trichinella spiralis* (Pennock & Grencis 2004). Loss of these cells from the bone marrow was followed by their appearance in the blood, with mature mast cells becoming detectable in the intestines three days later (Pennock & Grencis 2004).

It is thought that committed mast cell progenitors in the blood accumulate in tissues where they differentiate into mature mast cells. Their mature phenotype depends upon local factors produced in the tissue where they accumulate (Nakano et al 1985). The nature of these factors has been deduced from cell culture experiments that can mimic the stages of maturation. Studies with cultured mouse bone marrow cells have revealed some of the factors that induce mast cell maturation into specific phenotypes. Interestingly, FcεRI expression appears very early, with a significant proportion of the cells expressing the receptor after 1 week of culture with IL3, before the appearance of cytoplasmic granules (Thompson et al 1990). Several other factors can modulate differentiation and determine the mature phenotype e.g. IL4, IL9, IL10, M-CSF, GM-CSF, IFNα, IFNγ, NGF and SCF. Such soluble factors, in combination with signals picked up on interaction with local tissue cell surface or matrix molecules are believed to programme the maturation of mast cells in different locations. Based on studies in the mouse, two types of mature mast cell are recognized: mucosal and connective tissue-type. Mucosal mast cells are predominant in the lung and small intestinal mucosa, whereas connective tissue mast cells are found in skin, intestinal submucosa, blood vessel walls and heart. In the mouse, mucosal mast cells contain chondroitin sulfate and have a low histamine content, whereas connective tissue mast cells are longer-lived, contain heparin and have high levels of histamine. Mouse mast cells contain a number of serine proteases; mouse mast cell protease (mMCP)1 and mMCP2 being specific for the mucosal-type and mMCP4, mMCP5 and mMCP6 characterizing connective tissue-type mast cells (Miller & Pemberton 2002, Reynolds et al 1990).

Human mast cells are generally characterized according to their content of tryptase and chymase; cells associated with mucosal surfaces containing only tryptase (MC_T) and others, predominantly associated with connective tissue, containing both serine proteases (MC_{TC}) (Walls et al 1990). SCF is required in order to culture mast cells from human bone marrow or cord blood mononuclear cells and IL6 enhances the development of human mast cells from CD34^{+} cells in the presence of SCF (Saito et al 1996). The nasal epithelium and subepithelium of allergic rhinitis patients contain high numbers of MC_T, whilst nasal MC_{TC} are present in the deep lamina propria. MC_T preferentially express IL4, while MC_{TC} express IL-5 and IL-6 (Bradding et al 1995). Human MC_T and MC_{TC} both contain heparin, but MC_{TC} also have cathepsin G and carboxypeptidase A in their granules.

Mechanisms underlying the population of tissues with mast cells

The importance of mast cells in host defence to helminth parasites and allergic reactions is emphasised by the increase in tissue mast cell numbers seen in these situations. Thus, a marked mast cell hyperplasia occurs in tissues infected with parasitic helminths. Further, mast cell numbers can be 10-fold higher in the lungs of asthmatic patients (Gibson et al 1993) when compared to controls, and an increase in circulating mast cell progenitors has also been observed (Mwamtemi et al 2001). Patients with allergic rhinitis can have 50-fold more mast cells in the nasal mucosa in the pollen season (principally MC_T in the epithelium and sub-epithelium) when compared to the winter season (Fokkens et al 1992). These observations suggest that there are mechanisms that underlie the recruitment and local maturation of mast cell progenitors to maintain basal populations of mast cells in tissues, and also mechanisms to increase local mast cell numbers considerably in response to appropriate inflammatory signals.

One mechanism that has been explored to explain the localization of mast cells in specific tissues is via the expression of particular adhesion molecules on the surface of progenitors. Thus, it was demonstrated that the trafficking of mast cell progenitors to the small intestine was absent in β_7 integrin-deficient mice, while trafficking to the lung, spleen, bone marrow and large intestine was unaffected (Gurish et al 2001). Precedents from studies of other cell types suggest that cellular recruitment, either constitutive or induced, generally involves chemoattractant signalling molecules, acting in combination with adhesion molecules. Mast cells characteristically express c-kit and the ligand for this receptor, SCF, induces chemotaxis, in addition to other important effects such as proliferation, differentiation and inhibition of apoptosis. Several chemokine receptors have been identified on mast cells or mast cell lines e.g. CXCR2, CXCR3, CXCR4, CCR1, CCR3, CCR4 and CCR5 and their recognised chemokine ligands have been shown to be chemotactic for these cells. Of particular interest to us is the Eotaxin receptor CCR3 that is expressed on both mature human mast cells and their progenitors (Ochi et al 1999). Results obtained with CCR3 gene-deleted mice have been paradoxical. Thus, $CCR3^{-/-}$ mice infected with *T. spiralis* exhibited a normal jejunum and caecum mast cell hyperplasia and unaffected worm expulsion (Gurish et al 2002). However, $CCR3^{-/-}$ mice sensitized with ovalbumin intraperitoneally and challenged with aerosolized ovalbumin had increased numbers of tracheal intraepithelial mast cells and an increased hyperresponsiveness compared to wild-type mice (Humbles et al 2002). Recently, $CXCR2^{-/-}$ mice have been reported to have reduced numbers of intestinal mast cell progenitors as determined by limiting dilution assays, suggesting that mediators such as KC and MIP-2 are involved in trafficking whereas there was no reduction in mast cell progenitors in $CCR2^{-/-}$, $CCR3^{-/-}$ or $CCR5^{-/-}$ mice (Abonia et al 2005).

Mast cells exhibit distinct patterns of distribution within tissues. Thus, following mast cell progenitor recruitment via blood microvessels, mechanisms exist to relocate cells to extravascular sites, e.g. airway epithelium. This relocation is associated with phenotypic changes. Thus, mast cells that reside at different locations in the jejunum of mice infected with *T. spiralis* exhibit sequential changes in their granule ultrastructure and chymase phenotype (Friend et al 1996). Extravascular movement of mast cells may be dependent on chemoattractant gradients within tissues. Receptors on mast cells involved in recruitment from the blood could be down-regulated as maturation progresses in the tissue, to be replaced by receptors important for movement away from the blood microvessels where the cells are initially recruited. There is evidence that the precise location of mast cells within a tissue is important in determining their function. Thus, mast cells have been observed in airway smooth muscle in allergic asthmatic patients that have airway hyperresponsiveness to spasmogens, but not in patients with eosinophilic bronchitis where hyperresponsiveness is not a feature (Brightling et al 2002).

The 50-fold increase in mast cell numbers that can occur in the nasal mucosa of allergic rhinitis patients during the pollen season suggests that multiple exposure of the sensitized tissue to allergen results in the local production of chemoattractants for mast cell progenitors and an increased production of maturation factors. The chemoattractants then mediate progenitor recruitment from the blood. The same chemoattractants may be produced constitutively to mediate basal trafficking. Once in the general circulation these chemoattractants may also mediate the release of progenitors from the bone marrow, by analogy with other leukocyte types. With this as our working hypothesis, we have established two mouse models in order to identify candidate molecules with the ability to act as chemoattractants for mast cell progenitors.

An attractive mechanism to explain hyperplasia would be that mature mast cells, during activation, release a chemoattractant for progenitors. To address this, we cultured mouse femoral bone marrow cells with IL3 and tested them on 96 well chemotaxis plates with a range of chemotactic agents, using cells as immature as practicable. After 14 days in culture, cells were approximately 50% c-kit$^+$ with the major contaminating cells being Gr-1$^+$. The c-kit$^+$ cells were identified as mast cells by double labelling with other mast cell markers, FcεRI, CD34, T1/ST2 and CD13, as well as the α_4 and β_7 integrins. Mixed cell populations (Gr-1$^+$/c-kit$^+$) were tested on the chemotaxis plates. After a 3 hour incubation, migrated cells were counted by FACS analysis, counting the c-kit$^+$ and Gr-1$^+$ cells separately. Using this system on immature cells, we were surprised to find that the c-kit$^+$ cells (bone marrow-derived mast cells, BMMCs) were unresponsive to all the agents tested, i.e.

cytokines/growth factors, chemokines and peptides (Weller et al 2005). Despite their refractility to these molecules, immature mast cells were highly responsive in chemotaxis assays to supernatants of mature mast cells sensitised with IgE and activated by antigen. A single peak of chemotactic activity was found using reversed phase HPLC purification, coeluting with leukotriene B4 (LTB4) (Weller et al 2005). No other acute activation products were active at physiological concentrations in this system. Interestingly, responsiveness to LTB4 was lost as the cells matured, correlating with down-regulation of the BLT1 receptor (Weller et al 2005).

We tested LTB4 in chemotaxis assays using fresh bone marrow cell suspensions from mice. Migrated cells were cultured for 2 weeks in a medium containing TGFβ1 that favours rapid maturation to mucosal-type mast cells (Brown et al 2003). Cells were then lysed and mast cell-specific proteases mMCP1 and mMCP2 measured by ELISA. This amplification system allowed an indirect measurement to be made of the very few progenitor cells migrating from the bone marrow cell suspension. These experiments clearly demonstrated that mast cell progenitors in the bone marrow respond chemotactically to LTB4, but not to SCF. In comparison, c-kit$^+$ cells cultured for 2 weeks with IL3 still responded chemotactically to LTB4 , but also to SCF (with lower efficacy), suggesting that the c-kit receptor couples to the locomotor machinery during this period. Fluorescently-labelled 2 week c-kit$^+$ cells, when injected intravenously, were also able to accumulate in response to intradermally-injected LTB4 *in vivo* (Weller et al 2005). Human mast cells were cultured from umbilical cord blood using established techniques (Ochi et al 1999). These cultures demonstrated two distinct populations of c-kit^{high} and c-kit$^+$ cells, representing mature and immature mast cells respectively. In these experiments, the c-kit$^+$ cells responded chemotactically to LTB4 but not SCF, whereas the reverse was true for the c-kit^{high} cells (Weller et al 2005). This suggests that, as in the mouse, BLT1 is important in immature human mast cells and down-regulated as the cells mature.

In the future, we plan to explore in detail chemotactic responses of mast cells as they mature, responses that are important for the migration of the cells at different stages of their life history, and responses that would be expected to differ depending on the phenotype and localisation of the cells. Results so far suggest that BLT1 is important in the early stages, potentially to induce migration of progenitors through the bone marrow sinus endothelium into the blood and recruitment into tissues. Other receptors may have a role in subsequent relocalization within tissues on further maturation.

Conclusion

Tissue mast cells have an important role in innate and adaptive immunity. An understanding of the mechanisms involved in trafficking between compartments of the body will further knowledge about the life history of this cell and its interactions with other cell types. This may provide targets for therapeutic intervention.

Acknowledgements

We thank Asthma UK and the Wellcome Trust for supporting our research.

References

Abonia JP, Austen KF, Rollins BJ et al 2005 Constitutive homing of mast cell progenitors to the intestine depends on autologous expression of the chemokine receptor CXCR2. Blood 105:4308–4313

Applequist SE, Wallin RP, Ljunggren HG 2002 Variable expression of Toll-like receptor in murine innate and adaptive immune cell lines. Int Immunol 14:1065–1074

Arock M, Ross E, Lai-Kuen R, Averlant G, Gao Z, Abraham SN 1998 Phagocytic and tumor necrosis factor alpha response of human mast cells following exposure to gram-negative and gram-positive bacteria. Infect Immun 66:6030–6034

Bradding P, OkayamaY, Howarth PH, Church MK, Holgate ST 1995 Heterogeneity of human mast cells based on cytokine content. J Immunol 155:297–307

Brightling CE, Bradding P, Symon FA, Holgate ST, Wardlaw AJ, Pavord ID 2002 Mast-cell infiltration of airway smooth muscle in asthma. N Engl J Med 346:1699–1705

Brown JK, Knight PA, Wright SH, Thornton EM, Miller HR 2003 Constitutive secretion of the granule chymase mouse mast cell protease-1 and the chemokine, CCL2, by mucosal mast cell homologues. Clin Exp Allergy 33:132–146

Chen CC, Grimbaldeston MA, Tsai M, Weissman IL, Galli SJ 2005 Identification of mast cell progenitors in adult mice. Proc Natl Acad Sci USA 102:11408–11413

Di Nardo A, Vitiello A, Gallo RL 2003 Cutting edge: mast cell antimicrobial activity is mediated by expression of cathelicidin antimicrobial peptide. J Immunol 170:2274–2278

Echtenacher B, Männel DN, Hültner L 1996 Critical protective role of mast cells in a model of acute septic peritonitis. Nature 381:75–77

Edelson BT, Stricker TP, Li Z et al 2006 Novel collectin/C1q receptor mediates mast cell activation and innate immunity. Blood 107:143–150

Fokkens WJ, Godthelp T, Holm AF et al 1992 Dynamics of mast cells in the nasal mucosa of patients with allergic rhinitis and non-allergic controls: a biopsy study. Clin Exp Allergy 22:701–710

Friend DS, Ghildyal N, Austen KF, Gurish MF, Matsumoto R, Stevens RL 1996 Mast cells that reside at different locations in the jejunum of mice infected with Trichinella spiralis exhibit sequential changes in their granule ultrastructure and chymase phenotype. J Cell Biol 135:279–290

Gibson PG, Allen CJ, Yang JP et al 1993 Intraepithelial mast cells in allergic and nonallergic asthma. Assessment using bronchial brushings. Am Rev Respir Dis 148:80–86

Gurish MF, Humbles A, Tao H et al 2002 CCR3 is required for tissue eosinophilia and larval cytotoxicity after infection with Trichinella spiralis. J Immunol 168:5730–5736

Gurish MF, Tao H, Abonia JP et al 2001 Intestinal mast cell progenitors require CD49dbeta7 (alpha4beta7 integrin) for tissue-specific homing. J Exp Med 194:1243–1252

Humbles AA, Lu B, Friend DS et al 2002 The murine CCR3 receptor regulates both the role of eosinophils and mast cells in allergen-induced airway inflammation and hyperresponsiveness. Proc Natl Acad Sci USA 99:1479–1484

Jamur MC, Grodzki AC, Berenstein EH, Hamawy MM, Siraganian RP, Oliver C 2005 Identification and characterization of undifferentiated mast cells in mouse bone marrow. Blood 105:4282–4289

Kitamura Y, Go S, Hatanaka K 1978 Decrease of mast cells in W/Wv mice and their increase by bone marrow transplantation. Blood 52:447–452

Leal-Burumen I, Conlon P, Marshall JS 1994 Il-6 production by rat peritoneal mast cells is not necessarily preceded by histamine release and can be induced by bacterial lipopolysaccharide. J Immunol 152:5468–5476

Lin TJ, Maher LH, Gomi K, McCurdy JD, Garduno R, Marshall JS 2003 Selective early production of CCL20, or macrophage inflammatory protein 3alpha, by human mast cells in response to Pseudomonas aeruginosa. Infect Immun 71:365–373

Malaviya R, Ikeda T, Ross E, Abraham SN 1996 Mast cell modulation of neutrophil influx and bacterial clearance at sites of infection. Nature 381:77–80

Marshall JS, Jawdat DM 2004 Mast cells in innate immunity. J Allergy Clin Immunol 114:21–27

Miller HR, Pemberton AD 2002 Tissue-specific expression of mast cell granule serine proteinases and their role in inflammation in the lung and gut. Immunology 105: 375–390

Mwamtemi HH, Koike K, Kinoshita T et al 2001 An increase in circulating mast cell colony-forming cells in asthma. J Immunol 166:4672–4677

Nakano T, Sonoda T, Hayashi C et al 1985 Fate of bone marrow-derived cultured mast cells after intracutaneous, intraperitoneal, and intravenous transfer into genetically mast cell-deficient W/W^{v} mice: Evidence that cultured mast cells can give rise to both connective tissue type and mucosal mast cells. J Exp Med 162:1025–1043

Ochi H, Hirani WM, Yuan Q, Friend DS, Austen KF, Boyce JA 1999 T helper cell type 2 cytokine-mediated comitogenic responses and CCR3 expression during differentiation of human mast cells in vitro. J Exp Med 190:267–280

Pennock JL, Grencis RK 2004 In vivo exit of c-kit+/CD49d(hi)/beta7+ mucosal mast cell precursors from the bone marrow following infection with the intestinal nematode Trichinella spiralis. Blood 103:2655–2660

Reynolds DS, Stevens RL, Lane WS, Carr MH, Austen KF, Serafin WE 1990 Different mouse mast cell populations express various combinations of at least six distinct mast cell serine proteases. Proc Natl Acad Sci USA 87:3230–3234

Rodewald HR, Dessing M, Dvorak AM, Galli SJ 1996 Identification of a committed precursor for the mast cell lineage. Science 271:818–822

Saito H, Ebisawa M, Tachimoto H et al 1996 Selective growth of human mast cells induced by Steel factor, IL-6, and prostaglandin E2 from cord blood mononuclear cells. J Immunol 157:343–350

Sonoda T, Ohno T, Kitamura Y 1982 Concentration of mast-cell progenitors in bone marrow, spleen, and blood of mice determined by limiting dilution analysis. J Cell Physiol 112: 136–140

Supajatura V, Ushio H, Nakao A et al 2002 Differential responses of mast cell Toll-like receptors 2 and 4 in allergy and innate immunity. J Clin Invest 109:1351–1359

Thompson HL, Metcalfe DD, Kinet JP 1990 Early expression of high-affinity receptor for immunoglobulin E (Fc epsilon RI) during differentiation of mouse mast cells and human basophils. J Clin Invest 85:1227–1233

Walls AF, Bennett AR, McBride HM, Glennie MJ, Holgate ST, Church MK 1990 Production and characterization of monoclonal antibodies specific for human mast cell tryptase. Clin Exp Allergy 20:581–589

Weller CL, Collington SJ, Brown JK et al 2005 Leukotriene B4, an activation product of mast cells, is a chemoattractant for their progenitors. J Exp Med 201:1961–1971

DISCUSSION

Brown: Are mast cells the only source of leukotriene B4 (LTB4)?

Williams: No, mast cells are not the only source of LTB4. Mature mast cells secrete LTB4 during activation, but other cells also produce this lipid. For example, neutrophils secrete LTB4 (and IL8) during phagocytosis.

Finn: Presumably this is a steady-state migration population. How do things change in inflammation? Do they look for inflammatory sites in the epithelium?

Williams: There are two major parts to this. First, there is basal trafficking. Mast cells have a long life in the tissues of around 6 weeks. Continuous low-level replenishment of the cells is necessary. It may be that there is just a low level of LTB4 to replenish these cells, or there could be other signals. Second, and what we are really interested in, is mast cell hyperplasia. Here it looks as if activation of mast cells releases LTB4, and this is the mechanism used to recruit progenitors. Repeated exposure of tissues to allergens in sensitized individuals results in a marked mast cell hyperplasia and LTB4 may play a major role in this. We were hoping for something much more specific, because LTB4 also attracts neutrophils. However, having seen how the BLT1 receptor is expressed transiently in mast cells at a critical early stage in their life history, this leads us to believe that LTB4 has an important role in hyperplasia. Specificity may be conferred at a later stage. Recruited neutrophils have a short life in tissues. Mast cell progenitors will also have a short life in the absence of particular growth factors. If factors such as stem cell factor (SCF) are present in the tissues then mast cells will proliferate and mature.

Finn: It may be interesting to see what effect this increased production has on the endothelium.

Williams: I do not know of evidence that LTB4 can activate the microvascular endothelium directly. LTB4, given intradermally, induces the recruitment of immature mast cells from the blood. This is probably the result of integrin up-regulation induced by stimulation of the BLT1 receptor on the progenitor itself. When mature mast cells in tissues are activated they can release TNFα as well as LTB4. The TNFα can then stimulate the endothelium to increase expression of adhesion molecules. Thus, *in vivo* there may be synergism between stimulation of complementary adhesion molecules on the endothelium and the mast cell precursor.

E Sim: I wondered about Zileuton, a lipoxygenase inhibitor and also Montelukast, a receptor antagonist, that have recently been introduced for treating asthma with some success.

Williams: Montelukast would not affect the pathways involved in mast cell progenitor recruitment discussed here as it blocks peptidoleukotriene receptors. Zileuton, on the other hand would have an effect. As a 5-lipoxygenase inhibitor, it would block production of peptidoleukotrienes and LTB4. As an alternative, we are particularly interested in the clinical use of BLT1 antagonists as potential therapy for allergic diseases. Several companies have produced BLT1 antagonists in the past with the intention of using them in treatment for inflammatory diseases, such as rheumatoid arthritis. These were unsuccessful and the development programmes were discontinued.

E Sim: There is the interesting story of Opren which was used successfully for rheumatoid arthritis but which had side effects. I was surprised to see Zileuton introduced because Opren was a non-selective cyclo-oxygenase inhibitor and a lipoxygenase inhibitor. It has been difficult to get drugs of this type into the market because it was considered that the side effect, severe hepatotoxicity, was a pharmacological side effect and would always be the case. Zileuton shows that this isn't right.

Williams: I think BLT1 antagonists would be worth considering as an alternative to 5-LO inhibitors.

Lambrecht: As you know there has been quite a shake-up in the asthma field. The story is that the localization of mast cells is important: whether airways become hyperreactive or not isn't just a factor of the number of mast cells, but where they are in the tissues. Particularly when they are in the smooth muscle layer they might contribute to BHR. Could your pathway be involved in selective localization?

Williams: We have concentrated on the way the cells get from the microvasculature into the tissue. Peter Bradding has been looking at how they relocate once they are in the tissues. Their story is that it is a chemokine-driven effect.

Lambrecht: Do you get IP10 in your system?

Williams: No, it doesn't work on these immature cells and I don't think it is involved in recruitment of progenitors. What they are probably looking at is a movement of cells within the tissues. They say that IP10 is produced by the smooth muscle cells and CXCR3 is up-regulated on the mast cells in the tissues. This could be one of the following events, but we don't think it is involved in the recruitment from the blood.

Segal: What is in those granules? They look fantastic.

Williams: There is a whole range of serine proteases (which are characteristic of the different types of mast cell), highly-sulfated proteoglycans and histamine. In human, two main types of mast cell are recognized, but it is now thought that there is considerable plasticity with the cells changing their phenotype as they move between compartments.

Segal: What are those proteases doing?

Williams: There is a substantial literature on this. Some of them will be involved in the movement of the cells through tissues, and in tissue turnover as mast cells are implicated in remodelling, e.g. in wound healing. Others are involved in host defence reactions.

Mantovani: The lack of response to all chemokines is intriguing. Does this correspond to a lack of receptor expression, or are there receptors which are uncoupled?

Williams: We are now culturing the cells using different cytokine mixtures to produce either mucosal or connective tissue types, and then we are using a panel of antibodies to see which receptors are expressed. The chemokine receptor we have concentrated on is CCR3, but it is difficult to demonstrate this in the mouse cells. We see CCR3 mRNA, but we probably need activation of the cells to get the receptors expressed on the cell surface. So far we have seen very little. I was hoping that there would be a novel chemokine receptor on the very immature mast cells, but this is not the case.

Lambrecht: You chose c-kit and Gr-1 as a marker. c-kit is also found in stem cells and early myeloid progenitor cells. I suspect after two weeks of this assay the majority of stem cells are gone. How confident are you that there aren't any truly undifferentiated progenitor cells present? Your Gr-1 positivity is also recognizing Ly6C which is present on early monocytes.

Williams: We only use the Gr-1 antibody to eliminate cells in the cultures that are not immature mast cells. You are right: c-kit will be on earlier cells. I think there will be few of these left when we carry out the experiments with cultured cells. We do use a range of other markers too to identify the immature mast cells. In the situation where we are measuring chemotaxis in a fresh suspension of cells from bone marrow, it could be that we have got an even more primitive cell responding to the LTB4. We do not know the exact time at which the cells start to express BLT1. It could occur before commitment to the mast cell lineage.

Ryffel: Coming back to your two types of mast cells, there are two mouse models that lack mast cells, Sl (the WW^v mouse) and then the histamine decarboxylase (HDC) knockout. Is there a preferential distribution in the mast cell knockout? Can you induce any mast cell response?

Williams: There has been a lot of work done on those knockouts. There are infection models that become lethal if you delete the mast cells. For example, the caecal ligation and puncture model is lethal if you delete mast cells, but if mast cells are reconstituted the lethality is lost.

Ryffel: Is there a difference between WW^v and HDC knockout mice? It appears that not all mast cells are ablated in the HDC knockout mice.

Williams: There are not many mast cells left in the c-kit or SCF knockouts. The relative importance of mast cells in a particular mouse allergy model depends on the mouse strain and the immunisation and challenge protocol.

Lambrecht: It depends on whether you are using adjuvant or not. If you use a strong Th2 adjuvant, most models are mast cell independent. Under weak conditions the addition of IgE is needed, along with the release of mediators.

Gordon: What happened to the basophils?

Williams: Basophils are superficially similar to mast cells, they contain histamine and they have high affinity receptors to IgE. However, basophils circulate and are recruited as mature cells. Their precise role is not clear, but they are seen in high numbers in particular types of inflammatory responses, e.g. in response to tick infection. Basophils are a good source of Th2 cytokines so they may have a function in amplifying responses.

Lambrecht: Everyone tends to see mast cells as bad cells in allergic inflammation, but they could be good cells. Through their release of COX2 products and PGD2 they could be suppressive. In a parasite model, PGD2 suppresses inflammation. Couldn't mast cells protect from lethality in infection models by preventing overt inflammation or by inducing remodelling of tissues?

Williams: They might well do. Also, worm expulsion from the gut is mast cell dependent in some models. They also have the potential to down-regulate inflammation and are involved in wound healing.

Innate immunity and mucus structure and function

John K. Sheehan, Mehmet Kesimer and Raymond Pickles

Cystic Fibrosis Center, Campus Box 7248, 4019a Thurston Bowles, University of North Carolina at Chapel Hill, NC 27599, USA

Abstract. Many of the proteins associated with innate immunity in the upper respiratory tract are to be found localized into mucus gels and the mucin-rich surface layers of the epithelium and the cilia. Mucus is a relatively dilute suspension of such macromolecules being around 2–4% solids in normal induced sputum. These proteins scavenge, immobilise and/or kill pathogens and at the same time immobilize them into the mucus. Mucus is moved from the lung by the mucociliary clearance mechanisms or by cough. Some 190 proteins are readily detectable in sputum by proteomics methods and about 100 in bronchial air–liquid interface culture secretions. This cell culture system mimics the surface ciliated phenotype of the large airways very well and about 85 secreted proteins are common to both culture and sputum secretions. The major single protein by weight in cell culture secretions is MUC5B and in sputum a mixture of MUC5B and MUC5AC. The three epithelial mucins MUC1, 4 and 16 are also detectable in both secretions. In this paper the roles that these molecules play in protecting and stabilising the ciliated surface and building the gel will be discussed. The role of water and ion homeostasis is particularly crucial in mucus gel formation and evidence is gathering that it is perturbation of hydration mechanisms that may play into defective mucus leading subsequently to stasis and mechanical problems.

2006 Innate immunity to pulmonary infection. Wiley, Chichester (Novartis Foundation Symposium 279) p 155–169

Our lung is a particular miracle of evolution: it is simply not allowed to go out of commission. It has a surface area of a tennis court and is exposed to about 10 000 litres of air a day which is often contaminated with a staggering array of physical particles of different sizes and surface properties, biological organisms, toxic chemicals and noxious gases. Each generation makes up a new set of challenges for the system to overcome, the effluent from the internal combustion engine being an interesting case in point. In the main, even when grossly and gratuitously insulted by, for example, tobacco smoke, the lung functions quite effectively over a period of 50 years or so and, when well maintained, much longer.

How all this is accomplished is still a mystery in detail and in this paper I wish to propose that it is reliant on the integration of a network of proteins to build an effective mucus gel with properties consistent with removal, either by the ciliary escalator or by cough. Evidence from hypersecretory pathologies indicates that mucus stasis is a major contributor to exacerbations and it is clear that much more knowledge is required on the nature of mucus, how it moves by flow and cough and what goes wrong in disease. It is our working hypothesis that mucus in the upper airway is intelligent and optimized to cope with the local array of insults (Knowles & Boucher 2002). Information on these insults is fed back via the cilia to the different epithelial protection mechanisms. In recent years, with the advent of air–liquid interface cell culture methods, new experiments and new physical insights are emergent from many laboratories on structure and function. Some aspects of the efforts at the CF Center in Chapel Hill and in the broader UNC community will be summarized here.

Mucus in the upper respiratory tract: a compositional view

What is mucus? Normal induced mucus has a bio-solids (including ions) to liquid ratio of around 2–4%; in other words, it is about 96% water. A normal cell culture will rapidly build up a surface mucus that is similar in solid content (Thornton et al 2000). This points to the fact that water and ion homeostasis on the mucosal surface are the bedrock of the system. The 4% biomolecules are comprised of around 200 molecules or so in sputum and about half that number in cell culture secretions. Of these about 180 in sputum are readily detected in a shotgun proteomic scan. In Fig. 1 these molecules have been represented in different functional groups on the statistical basis of peptide recovery in the mass spectrometer. The major single group are those that might be associated with innate immunity and these are listed in Table 1. Many are found in culture and the main differences reflect the presence of immunoglobulins that would not be expected in the culture model. However in terms of the single major contributor to sputum, it is the mucins that top the list and these have been broken out of the innate immunity proteins as a special group. There are only five proteins: the epithelial mucins MUC1, MUC4, MUC16 and the large so called gel-forming mucins MUC5AC and MUC5B, that contribute to this list in both cell culture and sputum. Of these five, MUC5B is dramatically dominant in culture secretions accounting for about 70–80% of the total mucin. In sputum MUC5AC and 5B are present in similar proportions and account for 90% of the mucins. Thus there is an intriguing and distinctive difference between culture and sputum in terms of the presence of MUC5AC. Due to their compositional and structural dominance in respiratory mucus it is worth considering these molecules in more detail.

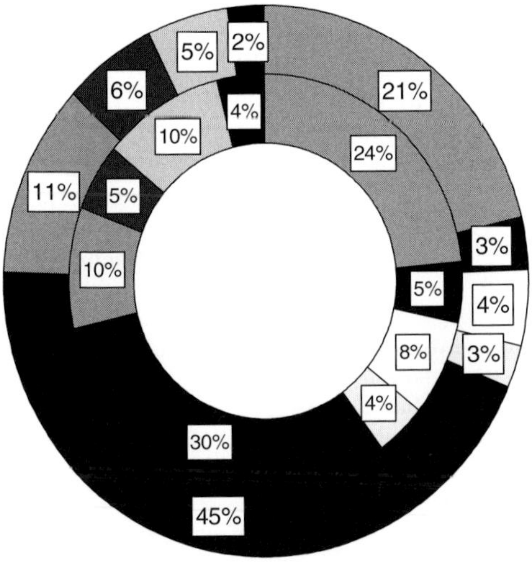

FIG. 1. All the peptides obtained by a mass spectrometry scan on cell culture secretions and human sputum have been put in nine selected protein categories. The outer ring is sputum and starting at 21/24% which are the mucins, the other eight categories going clockwise around the ring, are: antioxidant enzymes 3/5%, proteases and inhibitors 4/8%, other enzymes 3/4%, host defence and immune response 45/30%, cytoskeletal proteins 11/10%, calcium binding proteins 6/5%, membrane/signal proteins 5/10% and other 2/4%. The comparisons are based upon the number of peptides in the grouping (detected by mass spectrometry) as a fraction of the total peptides. The only grouping that was assessed by independent biochemical means was the mucin, which gave between 20–30% by weight for different preparations. No other quantitative data can be inferred from the above data but they do indicate a remarkable similarity and confirm the cell culture as an effective model system for many experiments concerning innate immunity.

The gel-forming mucin structure

A useful general reference to mucin structure will be found in Hollingsworth & Swanson (2004). A schematic structure of MUC5B and MUC4 that represent the two mucin families present in sputum is shown in Fig. 2. MUC5AC and 5B are large proteins in their own right and when fully glycosylated would have a molecular weight of $2–3 \times 10^6$ where 80% of that mass is protein. However they assemble into much larger oligomers that can reach molecular weights beyond 100×10^6. There is evidence for complex linear assemblies of these molecules but much remains to be done on their synthesis and secretion (Sheehan et al 2000). Their protein sequences and block structures are very similar and it is thought that the basic molecular mechanisms underpinning their assembly are similar. The MUC5AC mucin is sialic acid rich and is a rather homogeneous molecule with respect to glyosylation. It appears to be the major product of the surface goblet

TABLE 1 Mucus composition[a]

Macrophage migration inhibitory factor
Neutrophil gelatinase-associated lipocalin
Polymeric-immunoglobulin receptor
LPLUNC1/Von Ebner minor salivary gland protein
PLUNC/ SPLUNC1/Palate lung and nasal epithelium clone protein
Carcinoembryonic antigen-related cell adhesion
Deleted in malignant brain tumors 1/DMBT1 glycoprotein 340/gp340
Lysozyme
Lactotransferrin
Clara cell phospholipid-binding protein/Uteroglobin
Neutrophil defensin 1/Defensin, α1
Galectin-3 binding protein/Mac-2 binding protein
Leucine-rich α2-glycoprotein/LRG
CD59 glycoprotein/MAC-inhibitory protein
Clusterin/Apolipoprotein J, TRPM-2
β2-microglobulin
Dermcidin/Preproteolysin
Complement C3
Complement factor H/H factor 1
Zinc-<2-glycoprotein
Ig alpha-1 chain C region
Ig mu heavy chain disease protein
Ig kappa chain C region
Ig gamma-1 chain C region
Ig lambda chain C regions
Immunoglobulin J chain
Ig mu chain C region
Ig heavy chain V-III region TIL
Azurocidin/Cationic antimicrobial aptoglobin
α1-acid glycoprotein 1/Orosomucoid 1
Pulmonary surfactant-associated protein B
Pulmonary surfactant-associated protein A1
Trefoil factor 3/Intestinal trefoil factor
Haptoglobin

[a]This table lists a group of proteins detected by mass spectrometry associated with innate immunity. In bold (top) are proteins found in both cell culture secretions and sputum. Other molecules listed beneath were found in specific sputum samples.

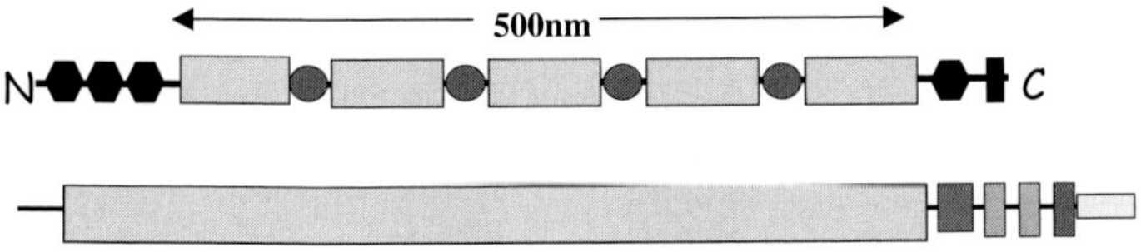

FIG. 2. A simplified schematic of the gel-forming MUC5B/5AC mucin structure is shown at the top. They are very similar in design. The most prominent region in both molecules is the large central domain indicated by the arrows that constitutes the heavily *O*-glycosylated region of the molecules. This domain has a small cys-rich peptide domain repeated at intervals. In MUC5B this whole domain is encoded by one 11.7 kb giant exon. When glycosylated this protein sequence dominates the physical properties of the molecule and is about 500 nm in length. The N and C termini have complex protein domain structure dominated by VWF D-domains. The mucins form large oligomeric structures through disulfide-mediated interactions in the termini. They oligomerize initially by C—C terminal interactions to generate dimers and subsequently through N–N interactions of these dimers. The lower simplified schematic is of an epithelial mucin such as MUC4. It contains a number of globular domain structures in the C-terminus and a small transmembrane sequence close to the extreme C-terminal end. The molecule is dominated by a very large highly glycosylated N-terminal domain that can be variable in length but be as long as 1–2 μm i.e. much longer than the gel-forming mucins above. MUC1 is similar but has a much shorter glycosylated domain (200 nm).

cells. The MUC5B mucin is undoubtedly a major product of the submucosal glands and is more heterogeneous, there often being distinct glycosylated forms present in sputum. The questions as to why functionally there are two mucins, and why they have different sites of synthesis and secretion and distinctive glycosylation patterns remain unanswered. Recently in a study of induced sputum we have found the MUC5B and MUC5AC proteins to be enriched in different and separable sputum phases (Fig. 3). The MUC5B mucin formed the basis of a very viscoelastic but dispersable fluid, whereas MUC5AC formed the basis of a viscoelastic gel that was separable from this fluid. These phases have some distinctive proteins associated with them and their obvious rheological differences suggest different functions. One idea under consideration is that they may form the basis of spatially distinct mucus gels one tuned to flow and the other to cough. This would embrace the idea of cough as an important natural mechanism of clearance of particulates tuned to surface goblet cell secretion.

Gel-forming mucin complexes

Clearly these two molecules play an important role in the formation of the mucus gel itself. There is considerable evidence that the mucins are assembled to their extreme size within the context of the cell (Sheehan et al 2000). Much remains to be done to understand the details of the mechanisms of synthesis and macromolecular assembly. Clearly the mucins may be maintained within the membrane

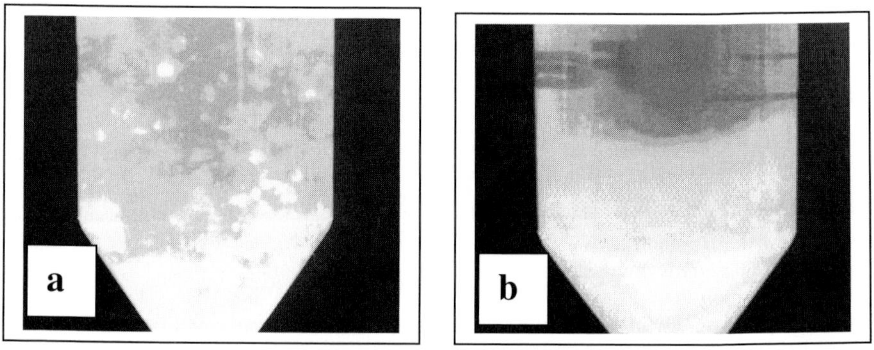

FIG. 3. Freshly induced sputum is shown diluted and dispersed into 10 volumes of saline (a). It is a viscoeleastic liquid containing translucent particulates. After centrifugation at 3500 g for 2 h the particulates fractionate to the bottom of the tube and form a coherent gel. This phase contains about 95% of the MUC5AC and 10% of the MUC5B while the fluid phase contains 95% of the MUC5B.

coated granules of the mucin secretory cells and are prompted to secrete by a variety of stimuli. The work of Verdugo suggests that upon secretion the mucin granules expand rapidly and fuse. Granule complexes which were isolated fresh from HT29 cells washed and imaged by electron microscopy are shown in Fig. 4a. The same preparation after partial purification by density gradient centrifugation is shown in Fig. 4b. There is a dense complex of proteins that sit in association with the filamentous mucins that can resist washing but are removable by gentle fractionation methods. What these proteins are and how they are organized with respect to the mucins is under investigation. The functional importance of globular proteins in contributing to mucus rheological properties is unknown.

Our proteomic studies in cell culture indicate that around 30 proteins sit in strong association with the mucins but we have little knowledge of the organizing principles underpinning these assemblies. It is possible to subfractionate them by a variety of physical methods and find two distinct regimes. One group appears to represent a subset of particles that emanate from the surface epithelium directly and are consistent in size, mass and density with membrane bound vesicles or perhaps exosomes. The proteins involved include actin, ezrin and villin and the epithelial mucin MUC1 which is strongly enriched in the microvilli on ciliated cells. On the other hand the gel-forming mucins are associated with a range of calcium binding proteins, antibacterial peptides and scavenging factors. There is, as yet, little or no information concerning the secretion of many of these proteins and how, where and by what mechanisms they come to be associated with the mucins.

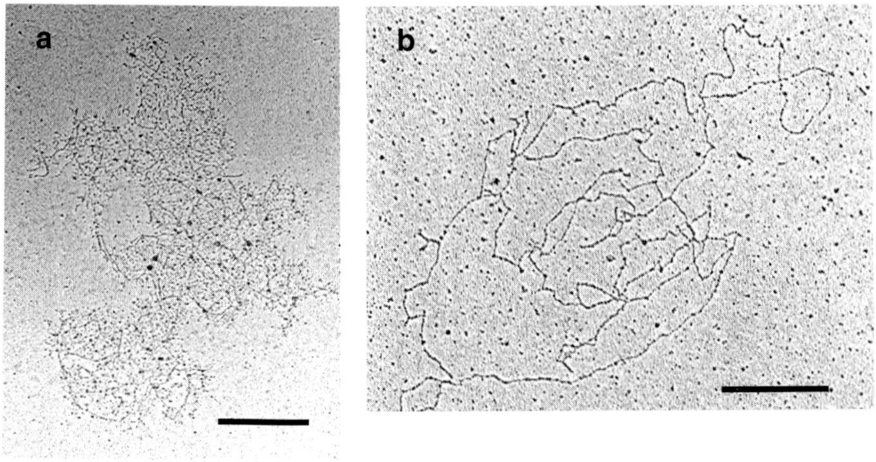

FIG. 4. Freshly secreted mucin complexes from HT-29 cells making MUC5AC were found to sediment rapidly at very low g forces (3000 g) and were recovered by this means. After re-suspension and washing in saline they were subjected to electron microscopy. The material was visualised as compact discrete islands of complex material (a). Subsequently this kind of preparation was subjected to isopycnic density gradient centrifugation in Cs2SO4 with a starting density of 1.35 gm/ml. The mucins isolated by this procedure (b) were found in discrete islands, similar in size to (a) but now clearly delineated. Much material was removed by the procedure. The bar is about 500 nm.

The epithelial mucins

MUC1, 4 and 16 all contribute to both cell culture and respiratory secretions. However, their abundance appears to be higher in culture secretions. A schematic diagram indicating the MUC1/4 domain structure is shown in Fig. 2. They all have multiple globular protein domains in their C-termini of unknown function and large glycosylated sequences towards their N-termini that dominate their physical properties. In MUC1 and 4 these sequences are polymorphic at the gene level and are found as variable numbers of tandem repeats determined by genetic factors which yield secreted glycoproteins of different sizes. The MUC16 mucin is very different. It has a very large and distinctive unique C-terminal domain dominated by multiple repeat motifs of a largely unglycosylated domain. N-terminal to this is a very large *O*-glycosylated domain. The total polypeptide core protein is estimated as being a gigantic 2200 kDa in size and the glycosylated protein we estimate as being 20×10^6 kDa. There is as yet, rather little information at the biochemical/biophysical level about these mucins. Histological studies on both tissue and cell culture sections indicate that MUC1 is an important contributor to the cell surface and is particularly rich around the microvilli of the ciliated cells. However, the cellular origin of these molecules is not clear, perhaps they are the products of all cells. On the other hand MUC4 is clearly an important product of ciliated cells, while intriguingly MUC16 at the intra-cellular level is associated with MUC5B secreting cells in both cell culture and in the glands in tissue section. However, all three of these molecules are found by microscopy to be in close association with the periciliary environment, whereas the secreted 5AC and 5B mucins appear to be rigorously excluded from this environment. MUC1, 4 and 16 are all to be found in culture secretions and washings and MUC1 and 4 particularly are part of protein complexes from the cell surface. These structures sometimes come complete with channels, enzymes and protective proteins but no clear functional role in mucus has been assigned to them.

Conjecture concerning the physical function for the epithelial mucins

The exclusion of the gel-forming mucins from around the cilia suggests that the epithelial mucins, particularly MUC1 and 4, might play a special role with regard to the ciliated cells. The current model that dominates the collective imagination envisages the cilia as surrounded by, and beating in, a low viscosity fluid. This layer around the cilia is often referred to as the pericilliary liquid layer (PCL). However, microscopy of this layer using methods that preserve structure indicate that it is more to be regarded as a dense negatively charged matrix (Fig. 5). The question is whether these mucins form a good basis for this layer and what the emergent properties would be. If it is imagined that the epithelial mucins connect to the cell only via their transmembrane domain and the extended carbohydrate-rich region is fully exposed to the hydrated extra-cellular milieu, then, from a polymer physics perspective, these mucins would be regarded as a grafted brush layer (Alexander 1997, DeGennes 1980). The properties of such layers have been described over many years and it is a prediction that as the density of the polymer increases so the layer will thicken as the chains stretch away from the surface (Fig. 6). If the inter-cilia distance is about 300 nm then opposing layers of surface located MUC4 might reasonably impinge on each other according to estimates of how large this molecule might be based upon protein sequence information. The theoretical physics of charged polyelectrolyte brushes indicates that little penetration would occur at the interfacial layer and experimental data confirm this and moreover further suggest that the interface would be extremely low in friction (Raviv et al 2003). Taken as a whole these data suggest that we should imagine the PCL more as a kind of 'gel' rather than a 'liquid layer'. A simple prediction from this view would be that much smaller particles than the 200–300 nm inter-cilia dimension would be excluded from this layer and indeed that is what is observed. Using polystyrene beads of different sizes it was found that the mesh size around the PCL is somewhat less than 25 nm and that particles beyond this size do not gain ready access to the surface epithelium. This is rather similar to the cell surface in rat gut

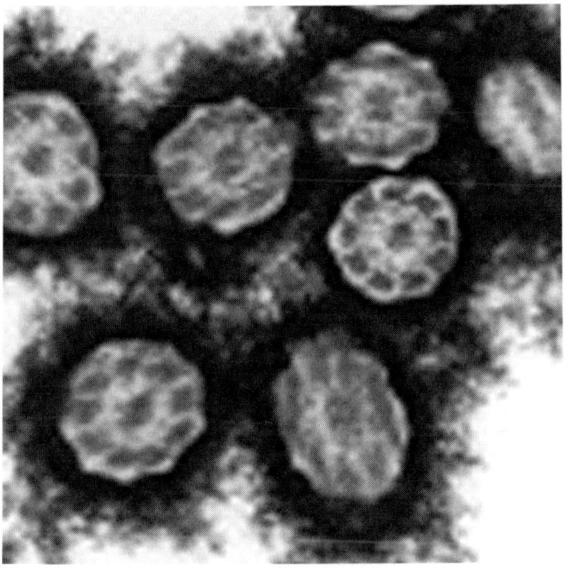

FIG. 5. Tracheal frozen sections cut in cross section were stained with ruthenium red. This stain binds very effectively to negatively charged molecules such as mucins. The presence of a dense charged network surrounding the cilia is evident. This is consistent with the presence of a dense coating of MUC4 epithelial mucins.

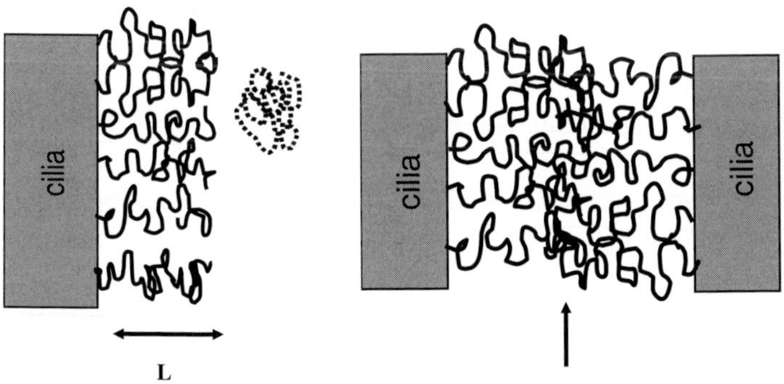

FIG. 6. The physics of the grafted brush concept as defined for cilia and their interactions is shown below. The diagram on the left demonstrates the concept that increasing the density of attached molecules past the point where they impinge upon each other, increases the thickness of the layer L due to interactions between the chains. As the layer extends away from the surface the pore size of the network shrinks and particles are excluded from the network on the basis of their size. On the right the physics of opposing such surfaces is demonstrated. At a critical surface polymer concentration the networks cannot interpenetrate and an interface is formed (see arrow). In the case where the polymers are polyelectrolytes the interface friction is dominated by the local chain-ion-water electrostatics. The net predicted result is that the interface will have very low friction. Compression of the volume between the cilia is predicted to be resisted and the energy stored and released by entropic effects of the polymer chains.

epithelium (Rambourg et al 1966). Another predicted property of these layers is that they would provide turgor that would maintain cilia organisation and coordination and store energy under constriction of the cilia that would occur at the stationary end points of the cilia beat cycle, this energy to be returned through the dynamic phase. Thus taken as a whole, this view of a dynamic brush is economic and has a number of advantages: the cilia are organized, maintained and the underlying epithelium protected in such a milieu while energy is conserved and friction minimized, all through the passive self-organization of these structural mucin components. Proving the details of such a conjecture is however difficult and presents many challenges not the least of which is the quantitation of the mucin components and our ability to manipulate them both in the test-tube and in their biological context.

Mucus and mucins in disease

Generally, from a clinical point of view mucus is viewed as a pathological entity. In many lung conditions morbidity and mortality is associated with hypersecretion of mucus and this is often clearly associated with hyperplasia and metaplasia of mucin-secreting cells, particularly goblet cells. However I would propose that it is not the amount of mucus that is produced that is so vital but the efficiency with which it is removed from the lung. The mechanisms for mucus removal are flow and cough. Even in individuals with ciliary dyskinesias mucus retrieval by cough alone is effective enough to yield a life expectancy, with the help of antibiotics, which is reasonable. Compare this with individuals having cystic fibrosis (CF), most of whom will have experienced severe lung problems through their childhood and teens and will be in a life-threatened state through their later life, many dying in their thirties. The question might be posed as to why individuals with CF cannot cough effectively and we do not have an answer. Clearly, blocking of the lung with tenacious, thick, sticky plaques that give home to bacteria is a dominating characteristic of this disease. There is increasing evidence from our work in cell culture that the CF phenotype in the uninfected state (attainable in culture) does not secrete a different compositionally distinctive mucus as compared with normal but that it does rapidly revert to an impeded flow phenotype. This cell culture phenotype projected into lung would suggest that individuals are subject to greater risk of plaque formation and subsequent infection and inflammation in early life. Over time this tendency to infection would promote an airway surface phenotype tuned to mucin hypersecretion thus exacerbating and intensifying the problem.

In advanced disease CF mucus is remarkable in its quality, quantity, physical properties and composition. The obvious characteristic is that its ratio of biomolecules and ions to water has risen to values closer to 20–30%. The dense bacterial conglomerates that colonize this mucus introduce new polymers into the system and there is certainly high molecular weight DNA present. However, the 5B and 5AC mucins are still the predominant molecules present but they have been greatly degraded by the action of bacterial proteases, and proteomic studies suggest that the other innate immunity molecules might share a similar fate. The rheology of this dense mixture is surprisingly more fluid-like than the 4% gels found in induced sputum. Is it possible that this scenario described here forms the common basis of much hypersecretory disease? Perhaps, from a global viewpoint the lung mucosal surface and the workings of the innate immune system are at the highest level reliant upon, and targeted towards, the optimization of flow. If this were to be so it would suggest a new functional emphasis and direction in studying its individual constituent proteins. The flow characteristics of the mucus and mucosal layer will be dominated by the mucin concentration and availability of water. This in turn is controlled by the ion transport mechanisms. It is the direct and indirect effects of infectious agents and environmental particulates and chemicals on these ion transport mechanisms which might have the most immediate impact on mucin network hydration. In turn it is possible that many specific proteins tasked with dealing with different insults will couple directly and perhaps indirectly via mucin association to secretion and water and ion transport.

Infection as effector of change in the epithelium

Combating infectious agents such as viruses and bacteria is probably the major business of the mucosal surface, and a number of the proteins associated with mucus either bind to and kill bacteria, immobilize them in the mucus network or target them for recognition by protective cells of the innate immune system. In most cases mucus flow might be the major method of removal of even live bacteria by their immobilization onto the carbohydrate framework of the mucus. However, some agents, especially viruses, might be small enough to reach and enter the epithelial surfaces. Studies of the direct effects of infectious factors on the mucus secretion phenotype are still in their infancy. Recent cell culture studies at UNC indicate that certain viruses (e.g. RSV) that target ciliated cells, can over some days through their infection cycle, engender large changes in surface morphology and secretory phenotype of the culture. The cultures sacrifice their ciliated cells and adopt a much more goblet cell-rich phenotype. The infected cells are expelled into a thick mucus layer that accretes on the cells. The fact that these responses are part of the structural phenotype of the cell culture surface itself and do not require addition of exogenous factors suggests that the proteins of the innate immune system may also play a key role in feedback mechanisms involved in the maintenance and re-modelling of the epithelial surface.

Some final thoughts

If some viruses alone can force a remodelling of the mucosal surface, what could the parallel or sequential exposure of the same biological surface to a range of viral, bacterial, particulate and chemical agents entail? One feels that the mucosal surface is trying to produce a network that may entrap, block or disable many different substances and at the same time maintain its dynamics. It is possible that different factors put different stresses on the system and require responses that compete in the optimisation process of the mucus engineering. Thus if more mucins are called for but there is not enough water for their hydration a sub-optimal mucus will result. In chronic pathological situations like CF, asthma and chronic obstructive pulmonary disease (COPD) the ultimate fall-back situation appears to be one of enhanced mucin and mucus production. This is a tricky balance to maintain and if there are new factors that prejudice hydration leading to plaques that cannot be effectively cleared, then the lung is put at risk. Adding something like smoking into the equation above when other unavoidable infectious and environmental irritants are prevalent in the environment could be seen as putting unrealistic engineering demands on the system. A prediction might be that smoking taken up by smoking-naïve populations subjected to numerous environmental and infectious factors as is the case in many deprived environments could have dramatic medium term effects on general respiratory function and increased penetrance of infectious agents.

References

Alexander S 1977 Adsorption of chain molecules with a polar head a scaling description. J Phys 38:983–989

De Gennes PG 1987 Polymers at an interface: a simplified view. Adv Colloid Interface Sci 27:189–209

Hollingsworth MA, Swanson BJ 2004 Mucins in cancer: protection and control of the cell surface. Nat Rev Cancer 4:45–60

Knowles MR, Boucher RC 2002 Mucus clearance as a primary innate defense mechanism for mammalian airways. J Clin Invest 109:571–577

Rambourg A, Neutra M, Leblond CP 1966 Presence of a 'cell coat' rich in carbohydrate at the surface of cells in the rat. Anat Rec 154:41–71

Raviv U, Glasson S, Kampf N, Gohy JF, Jerome R, Klein J 2003 Lubrication by charged polymers. Nature 425:164–165

Sheehan JK, Brazeau C, Kutay S et al 2000 Physical characterization of the MUC5AC mucin: a highly oligomeric glycoprotein whether isolated from cell culture or in vivo from respiratory mucous secretions. Biochem J 347 Pt 1:37–44

Thornton DJ, Gray T, Nettesheim P, Howard M, Koo JS, Sheehan JK 2000 Characterization of the mucins from cultured normal human tracheobronchial epithelial cells. Am J Physiol Lung Cell Mol Physiol 278:L1118–1128

DISCUSSION

Bateman: When one uses a hypertonic solution *in vivo*, does this just trigger cough, or does it in any way change the properties of the mucus?

Sheehan: There's a recent paper by a clinical group at Chapel Hill (Donaldson et al 2006) who are using hypertonic saline as a therapy for cystic fibrosis (CF). The major effect that they are exploiting is that if you put hypertonic saline onto the airway, it draws more water out of the cells. This helps loosen and perhaps move the mucus. I am doing it using myself as a model system. I am not sure how they have tuned the saline conditions to get the beneficial effect without getting what might be seen as a bad effect. I don't know how normal I am, though: I have never been a smoker but I do suffer from asthma. There may be some complexity here. The point of your question is that there will be a change in the nature and quality of the mucus that is made, and also in one's ability to clear it.

Bateman: We have been living with the convenient paradigm that there is a liquid (sol) layer adjacent to the cell surface the osmolarity of which can be easily altered by inhalation of hypertonic saline. Are you now saying that this periciliary layer of fluid does not exist?

Sheehan: I am putting something there, but these are gels. We are at 2% solids. In CF, for example, the concentration (biosolids, including ions) could get up to 14%. I once dragged material out of an asthmatic lung that was over 20% by weight mucin and other molecules. It was a rubber-like material. Once you get over 8% solids the mucus is too thick to be moved by cilia. It is also too thick for other things to move in it, which becomes important. A functioning gel appears to be around 1–4%, depending on conditions. It would be similarly so for this gel around the cilia. It is dominated by ions and water.

Schoub: Is the virus effect you described specific?

Sheehan: These are Ray Pickles's data in UNC Chapel Hill. He has shown me data for respiratory syncytial virus (RSV) and parainfluenza virus (PIV). The viruses get into the ciliated cells and the ciliated cells subsequently, over days, eject themselves out of the epithelium. The epithelium also subsequently adopts a mucus-rich cell phenotype over a period of days after infection. All the ejected ciliated cells, many still with their viral load, are seen in the mucus on top of the culture.

Speert: When you have a CF airway, where do biofilms form?

Sheehan: That's an interesting question. In CF, with this thickening of the mucus going up into the 10% solids regime, the bacteria change their phenotype. They can no longer be motile in this environment. This is an important effect: neutrophils might also change their phenotype if they can get into this environment. Again, it is this mechanoconstriction. What happens is that you are making a more anoxic environment which pulls in organisms that are more anaerobic; the pseudomonas lose their pilli and their motility, forming an aggregating phenotype. The film actually forms in the mucus.

Peiris: How do viruses get through this formidable barrier?

Sheehan: For me this is the $100 000 question. If we could solve this we would have a much better chance of doing gene therapy. A lot of gene therapy came from trying to arrange a vector inside the virus. We have always failed to transfer the virus across a healthy epithelium. In the experiments researchers pre-wash the cultures, taking away the mucus. What happens to virus on top of healthy mucus is an interesting question.

Peiris: But when the virus gets into a human or animal, it infects the cells through this mucus barrier.

Sheehan: My point is that it gets into specific cells. This is another crucial aspect. There is a lovely recent paper by Richard Grencis (Cliffe et al 2005) who studies parasites in the gut. He has developed a beautiful sacrificial model: the parasites get into the microvilli and these microvilli are all shed. Everything ends up in the mucus and new villi grow underneath. In virus experiments they get in, and then they get wrapped up. The whole ciliated cell, with its viral load developing, is expelled. This occurs at the same time as the formation of a sialic acid-rich mucus.

Lambrecht: I want to return to the point of uptake. We did some *in vivo* imaging with dual photon microscopy in which we looked at genetically tagged dendritic cells (DCs). Some DCs crawl up all the way underneath the tight junction, and then extend long dendrites into the airway lumen. They must cross this layer. Could these cylindrical structures which you see be the pathway through for dendrite extensions?

Sheehan: That's a brilliant thought: anything like this could be possible. One thing I didn't say is that many of the proteins found in the mucus come from the surface epithelium. Epithelial shedding is a basic mechanism by which the system is cleaning itself. What is found is these vesicular structures bleeding away from the system the whole time. They are wrapped up in vesicular form with MUC1. I think they play critical roles in bringing a host of proteins to bear up there in the mucus, or they are carrying stuff away from the surface epithelium and locking it away, so it can be put on the escalator.

Feldman: One of the other fallacies of the gel–sol layer hypothesis is that the mucus is a continuous blanket over the epithelium. If it was like this, with all these polymers, the cilia wouldn't budge it. Mucus moves on the airway in rafts. In most people there isn't all that much mucus there. Bacteria and viruses could just penetrate directly through to the naked epithelium.

Sheehan: That is not quite true. Over the tops of the cilia there are a lot of mucins condensed onto that structure. The pericilliary liquid layer (PCL) is not just the PCL. You are right about the raft of mucus that comes later, but over the cilia there is another structure. This has been described in the literature for a long time by electronmicroscopy, but I have recently been studying this by live cell microscopy. It is not quite the same structure as people have imagined on the basis of the electron micrographs.

Bateman: I would see the mucus rafts you described as rescue rafts for damaged or devitalized areas of the airway epithelium. We have many situations, such as in asthma, where the shedding of cells is a consequence of inflammation. The presence of rafts of mucus may be a dynamic protective mechanism to exposed mucosa.

Sheehan: For most of us living in the real world the lung has got plenty of work to do most of the time. This removal process is always being prompted. I suspect the raft is always in operation, and that is why we have our glands lubricating that surface.

van Helden: If we imagine that this raft of mucus is keeping the cell surface clear, then what is the future for aerosolized drugs, particularly those administered through nanoparticles?

Sheehan: There must have been billions of pounds spent trying to get drugs to cross the airway. There are very few drugs that do this effectively. I once sat in an aeroplane next to someone from GSK, and he said 'If you are able to invent a way of moving drugs easily and cheaply across the airway, you'll become a billionaire'. This is a barrier that has been designed to stop transfer, particularly trivial transfer of DNA. From an evolutionary perspective we are looking at a huge number of interlocking mechanisms to stop trivial transfer of anything that we might want to get across that barrier.

Brown: I have what may be a naïve question. If I understood correctly what you described, this PCL layer has cilia and carbohydrate tightly packed. How do the cilia beat, then?

Sheehan: Imagine two cilia. They are coated with the epithelial mucin MUC4. The microvilli are coated at another scale with MUC1, but the principle is the same. These epithelial mucins coating the cilia surfaces form a structure that maintains the spatial organization of the cilia. The physics of such coated surfaces predicts that they will form almost frictionless interfaces between the cilia allowing them to slide over each other freely. If something gets in between the cilia that binds to them, the cilia know about it since it will introduce friction into the system.

References

Donaldson SH, Bennett WD, Xeman KL, Knowles MR, Boucher RC 2006 Mucus clearance and lung function in CF with hypertonic saline. N Eng J Med 354:241–250

Cliffe LJ, Humphreys NE, Lane TE, Potten CS, Booth C, Grencis RK 2005 Accelerated intestinal epithelial cell turnover: a new mechanism of parasite expulsion. Science 308:1463–1465

Collectins and host defence

R. B. Sim*, H. Clark*, K. Hajela*† and K. R. Mayilyan*‡

**MRC Immunochemistry Unit, Department of Biochemistry, University of Oxford, South Parks Rd, Oxford OX1 3QU, UK, †School of Life Sciences, Devi Ahilya University, Indore 452001, India, and ‡Institute of Molecular Biology, Armenian National Academy of Sciences, 375014 Yerevan, Armenia*

Abstract. The collectins are a small family of soluble oligomeric proteins containing collagenous regions and C-type lectin domains. They are related in structure and function to complement protein C1q, and to H-, L- and M-ficolins. In humans, the collectins mannose-binding lectin (MBL) and surfactant proteins A and D (SP-A, SP-D) have important roles in innate immunity. MBL occurs mainly in blood plasma and in the upper respiratory tract. It binds to neutral sugar arrays on microorganisms and acts as an opsonin either directly (by binding to cell-surface calreticulin) or indirectly by activating complement. MBL circulates in complex with any of three proteases, named MBL-associated serine proteases (MASPs)-1, -2 and -3. MBL–MASP-2 complexes activate complement, but the role of MBL–MASP-1 and MBL–MASP-3 complexes is not yet known. MBL deficiency occurs at high frequency, and is associated with susceptibility to infection, particularly in infants. SP-A and SP-D are most abundant in the lungs, and also bind to microorganisms and inhaled particulates, mainly by lectin–sugar interactions. They do not activate complement, but act as opsonins and agglutinators, and have additional effects on cellular regulation. Mice deficient in SP-A or SP-D are susceptible to lung infections, and SP-D-deficient mice develop an emphysema-like condition.

2006 Innate immunity to pulmonary infection. Wiley, Chichester (Novartis Foundation Symposium 279) p 170–186

The proteins mannan-binding lectin (MBL), bovine conglutinin (BK) and lung surfactant proteins SP-A and SP-D were given the group name 'collectins' in 1992, in order to emphasize their emerging common properties in innate immunity (Malhotra et al 1992). Collectins contain polypeptide chains with a collagenous region and a C-type lectin domain. The collectin family remains small, and has expanded to include only two additional bovine collectins CL-43, CL-46 and two further human collectins CL-L1 and CL-P1 (Holmskov et al 2003, Hickling et al 2004). The collectins are a subset of a larger group of proteins, sometimes called 'defence collagens', which are proteins made up of polypeptides containing a collagenous region and, instead of a lectin domain, another type of globular domain. These include the complement protein C1q, the H-, L- and M-ficolins, and, perhaps, a number of recently characterized proteins of the C1q/TNF family (Tang et al

2005). The collectin MBL activates the complement system, as do C1q and the ficolins. Other collectins do not activate complement.

A major biological role of the collectins is to bind to targets (such as microorganisms [Table 1] or altered host cells) by recognizing patterns of carbohydrate distribution, and to enhance phagocytosis/clearance of the target. The lectin domain of each collectin binds to a range of monosaccharides (with equatorial hydroxyl groups at the 3rd and 4th carbon atoms: mannose, glucose and their amine forms, and also fucose), giving each of the collectins a potentially broad range of specificities (Holmskov et al 1994). The binding of collectins to carbohydrate ligands on the surface of targets elicits effector functions via the collagenous region. These effector functions include complement activation (for MBL) or recognition of the collectins by cellular receptors. The collectins that are involved in respiratory system defence are surfactant proteins A and D (SP-A and SP-D) and MBL (Hickling et al 2004).

Structures of collectins

The collectins are large oligomeric proteins, each composed of identical or very similar polypeptides (Fig. 1). Each polypeptide has four domains or regions: a cysteine-containing N-terminal portion, followed by collagen-like sequence, α-helical coiled-coil neck and, at the C-terminus, a C-type lectin domain (Fig. 1).

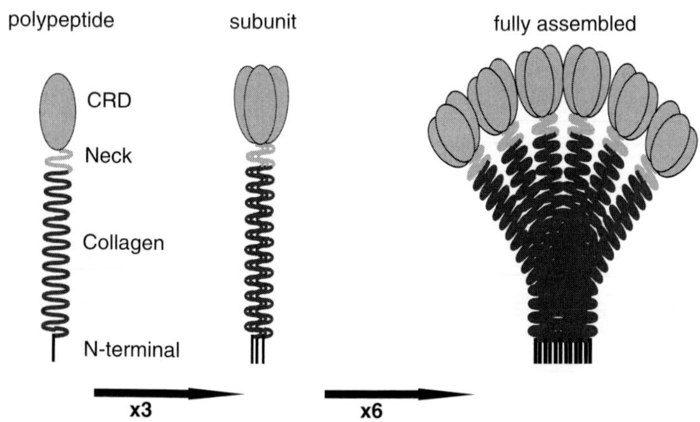

FIG. 1. The assembly of the collectins. As described in the text, groups of three polypeptide chains trimerize to form a single subunit with a 'stalk' made up of a collagen triple helix, and a 'head' made up of three lectin domains. The subunits assemble to form the full-size native collectins. The single polypeptide is made up of an N-terminal cysteine-containing region, a region of collagenous sequence, a coiled-coil sequence 'neck' and a CRD (carbohydrate recognition domain or lectin domain).

Three polypeptide chains form an elongated subunit, which has a 'head' composed of three C-type lectin domains, and a 'stalk', which is a collagen triple helix (Fig. 2). The subunits are linked into oligomers of up to six subunits, either covalently through disulfide bonding, or non-covalently. SP-A and MBL have the 'bunch-of-tulips' (sertiform) shape first characterized by electron microscopy for the complement protein C1q. SP-D has four subunits in a cruciform shape. These are very large proteins. Each subunit has a length of 46 nm in SP-D, 20 nm in SP-A and 13 nm in MBL. Polymerization is variable, particularly for SP-A, but the presumed major oligomer of each (human) collectin consists of six subunits (i.e. 18 polypeptides) for SP-A and four subunits for SP-D. Human MBL is likely to be mainly in a six subunit form, confirmed by hydrodynamic studies on recombinant MBL (Larsen et al 2004), but this is controversial, as Teillet et al (2005) suggest the commonest forms in plasma contain 3–4 subunits. The relative *in vivo* concentrations of the oligomeric forms of each collectin are not known accurately and are difficult to measure. The oligomerization is crucial for high avidity binding: the crystal structure of rat MBL showed that the individual carbohydrate-binding site of one C-type lectin domain forms only a very small, low affinity contact area with a single sugar (e.g. mannose) bound via a Ca^{2+} ion, so that multiple interactions of

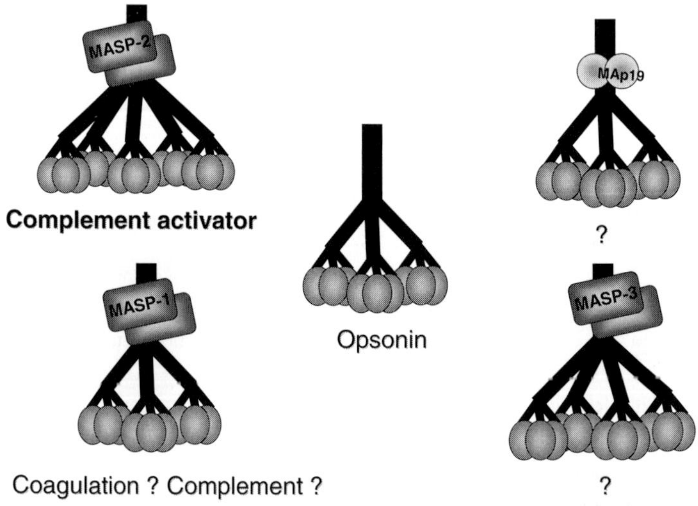

FIG. 2. The heterogeneity of MBL complexes with MASPs. As well as having several different oligomeric forms, MBL molecules may associate (or not) with any one of the homodimeric MASPs or with MAp19. Each complex may have a distinct function: the MBL-MASP-2 complex activates complement, and the 'naked' MBL is likely to act as an opsonin. The functions of the other complexes are unknown. The ficolins probably form a similar range of complexes with the MASPs.

several C-type lectin domains with a sugar array are needed for high avidity binding of the protein oligomer to a target surface (Weis & Drickamer 1994). The three sugar binding sites in a single subunit are about 5 nm apart. Therefore MBL and the other collectins can selectively recognize microorganisms, as opposed to host, by binding avidly only to large repetitive sugar arrays with appropriate ligand spacing.

The quaternary structures of SP-A and MBL are very similar to that of the complement component C1q, and to H, L and M-ficolin. It was partly the resemblance in shape of C1q and SP-A that stimulated research to show that SP-A, and other collectins, have, like C1q, a major role in innate immunity (Holmskov et al 1994, 2003). C1q is not a collectin, as its 'head' region consists of 'gC1q' domains that recognize charge motifs (as found on immunoglobulins in immune complexes, anionic phospholipids or on exposed lipid A on bacteria). The ficolins are also oligomers of polypeptides containing a collagenous region and a globular portion: in this case a fibrinogen-like domain (Akaiwa et al 1999, Holmskov et al 2003, Liu et al 2005). The binding specificities of the ficolins are not yet understood. All three ficolins (H, L and M) are loosely regarded as lectins, as they have been reported to bind *N*-acetyl amino sugars, such as GlcNAc and GalNAc, but Krarup et al (2004) suggest that the specificity of L-ficolin is simply for acetyl groups.

MBL and the ficolins

MBL (mannan-binding lectin) is also known as mannan-binding protein (MBP) or mannose-binding lectin, since it binds to yeast mannan as well as to mannose coupled to Sepharose. MBL is made in the liver, and is most abundant in blood, but is present in most body fluids, e.g. in buccal cavity and upper airway secretions, saliva and bronchoalveolar lavage fluid (Holmskov et al 2003, Presanis et al 2003). MBL concentration in plasma is very variable, ranging from 0 to over 20 μg/ml in humans (average 1–2 μg/ml) (Mayilyan et al 2006). MBL deficiency is common (5% or more of the population, depending on the threshold concentration used to define deficiency) and is associated with severe or repeated infections in infants (Super et al 1989) and with a wide range of infections and immune system-associated disorders in adults (Turner & Hamvas 2000). The variation in MBL levels can be attributed to three structural mutations of the MBL gene, which are likely to result in defective polymerization, interacting with several polymorphisms in the MBL promoter region that influence level of expression. The structural mutations occur at high frequency (generally 15% or greater cumulative allele frequency in most populations studied) and are single base changes resulting in amino acid changes in the collagen region of MBL, and in an altered capacity to form the collagen triple helix (Madsen et al 1998, Larsen et al 2004). The concentration and biochemical forms of MBL present in the airways are currently not known.

Three ficolins have been identified (L, H and M). All are oligomeric proteins similar in quaternary structure to MBL, but containing fibrinogen-like domains instead of C-type lectin domains (Akaiwa et al 1999, Holmskov et al 2003, Liu et al 2005). The binding specificity of the fibrinogen-like domains is not well established. All three ficolins activate the complement system, by binding the MASPs, the same proteases with which MBL associates. L-ficolin is found mainly in blood, but H-ficolin is also present in the lungs, made in Type II alveolar cells and ciliated bronchial epithelial cells. M-ficolin, which occurs mainly as a membrane-bound protein, also has a secreted form. It has been identified in secretory granules in Type II alveolar cells. The plasma levels of L- and H-ficolins are estimated at 4–14 μg/ml, and 7–23 μg/ml respectively, while M-ficolin is present in trace amounts. The concentrations of ficolins in the respiratory tract are unknown.

MBL, ficolins and complement

MBL can act directly as an opsonin, by binding to carbohydrates on pathogens, then interacting with MBL receptors on phagocytic cells. However, it can also trigger the opsonic activity of complement, resulting in deposition of C3b/iC3b on targets, and stimulation of phagocytic uptake via the C3 receptors, CR1, CR3, and CR4. MBL is the only one of the collectin family of proteins to activate the complement system: it shares this activity with C1q, and with the H-, L- and M-ficolins.

The complement system is a major mediator of innate immune defence and contributes to inflammation, opsonization and lysis. It consists of more than thirty proteins in plasma or bound to cell membranes. The complement system can be activated via three pathways, the classical, lectin or alternative upon recognition of pathogen-associated molecular patterns (PAMPs). Currently it is not clear whether complement activation has an important role in the respiratory tract, as in healthy individuals, the concentrations of complement proteins are very low in body fluids other than blood plasma. MBL and the ficolins form Ca^{2+} ion-dependent complexes with three proteases, named MBL-associated serine proteases (MASP-1, MASP-2 and MASP-3) (Petersen et al 2001, Hajela et al 2002). The MASPs are homologues of the complement classical pathway proteases, C1r and C1s, which bind to the complement protein C1q. The MASPs are activated from proenzymic to active form when MBL or ficolins bind to a target. MASP-2 activates the complement proteins C2 and C4, so is responsible for further complement activation. MASP-1 does not directly activate the complement system although earlier reports in the primary literature, and currently in reviews, suggest that it activates C3. This confusion may arise from the observation that it does cleave $C3(H_2O)$, an inactived form of C3, but it does not cleave active C3 as found in plasma. MASP-1 may be involved indirectly in C3 turnover (e.g. by activation of another protease) but this has not been investigated. MASP-1 will cleave fibrinogen and coagulation factor XIII *in vitro*, so may have a role in localised coagulation, though it is not clear if these are physiological substrates. No substrate has yet been found for MASP-3.

In contrast to the C1q complex formed with the MASP homologues C1r and C1s, MBL complexes with MASPs appear to be a very heterogeneous population. The C1q-$C1r_2$-$C1s_2$ complex has a fixed stoichiometry, and C1q has only a single oligomeric form (six subunits: a minor 2-subunit form does not bind C1r and C1s). MBL has variable oligomerization, and oligomers with 2 or more subunits probably only bind one type of MASP at a time. The MASPs exist as homodimers, so separate populations of MBL-$(MASP\text{-}1)_2$, MBL-$(MASP\text{-}2)_2$ and MBL-$(MASP\text{-}3)_2$ are likely to be in circulation. A truncated alternative splice product (MAp19) of the MASP-2 gene is also in circulation, and will form MBL-$(MAp19)_2$ complexes (Fig. 2). MAp19 has no proteolytic activity as it lacks a serine protease domain. The MASPs are also shared by the ficolins, and current estimates of MASP, MBL and ficolin concentrations suggest that there will be an excess of free MBL and ficolins, not complexed to MASPs or MAp19. Studies of MASP-1 and MASP-2 bound to MBL in individual human sera show that there is inverse correlation between MASP-1 and MASP-2 (i.e. on average, the more MASP-1 is bound to MBL, the less MASP-2 is bound) (Fig. 3) (Mayilyan et al 2006).

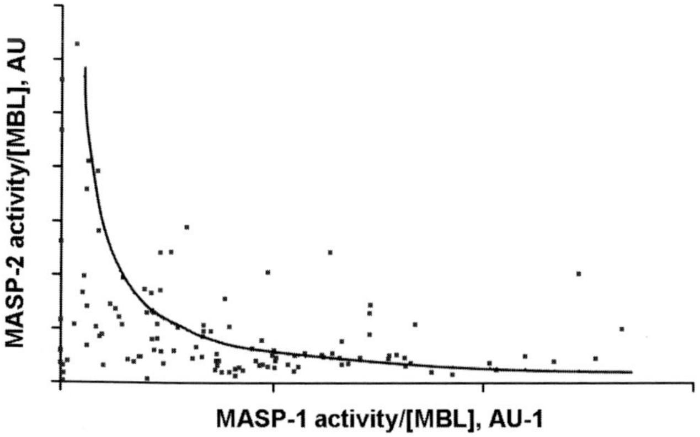

FIG. 3. The inverse relationship between MBL-bound MASP-1 and MBL-bound MASP-2. Assays were done to bind MBL and MASPs from human sera onto mannan-coated plates. The quantity of bound MBL, and activity of the bound MASP-1 and MASP-2 are measured, then MASP-1 and MASP-2 activity expressed (in arbitrary units) per unit of MBL. As discussed in the text, studies of large numbers of individual human sera show that MASP-1 and MASP-2 are inversely correlated, unlike the homologous proteases C1r and C1s of the complement classical pathway.

The surfactant proteins A and D

SP-A and SP-D are synthesized by alveolar type II and Clara cells in the lung and are most abundant in lung surfactant. Both are also found on mucosal surfaces outside the lung (Holmskov et al 1994, 2003, Kishore et al 2006). These proteins function in innate immunity by recognition and binding to non-self or altered self, and promoting clearance/phagocytosis by opsonization or agglutination. Additional roles in cell stimulation, not involving target adhesion to cells, are becoming apparent (Clark et al 2003, 2004). In the lung, SP-D is soluble, while SP-A is mainly intercalated into the tubular myelin layer.

SP-A and SP-D probably exist *in vivo* as a mixture of polymers. SP-A in lung lavage fluid is present in forms with 1 to 6 collectin subunits. The relative quantity of each oligomer varies between individuals and may be related to lung disease (Hickling et al 1998). As noted above, the binding affinity of a single lectin domain for carbohydrate is very low and the greater multiplicity of lectin domains found in higher-order multimers of SP-A, MBL and SP-D is required to give high-avidity binding to carbohydrate bearing surfaces. Isolated SP-A is mainly hexameric in structure. The mechanisms controlling the degree of polymerization are unclear, but stable oligomers depend on disulphide bridging. Although the factors controlling oligomer size are poorly understood, it may be influenced by redox changes in the lung. Depolymerization would be expected to lead to the loss of binding affinity for carbohydrate-rich surfaces with loss or alteration of biological function.

Activities of SP-A and SP-D

The biological functions of SP-A and SP-D include surfactant homeostasis (Hawgood & Poulain 2001) and host defence (Clark et al 2004, Kishore et al 2006). This review focuses on the host defence functions. SP-A and SP-D bind and agglutinate microorganisms and other particulate material entering the lungs. The full spectrum of *in vivo* targets of SP-A and SP-D has not been systematically investigated. However, many pathogens to which the surfactant associated collectins bind have been identified (Table 1) (Hickling et al 2004, Kishore et al 2006). In addition to binding respiratory pathogens, the collectins also bind allergenic particles, including house dust mite extracts and pollen grain granules (Malhotra et al 1993).

SP-A and SP-D also appear to modulate local inflammatory and immune responses. The collectins have been shown to promote attachment, uptake and killing of respiratory pathogens by alveolar macrophage. SP-A is essential for host defence as SP-A knockout mice are susceptible to infection from bacteria, viruses and fungi (Kishore et al 2006). SP-A and SP-D have a more complex role in

TABLE 1 Microorganisms to which collectins have been reported to bind

	Collectin		
Micro-organism	*MBL*	*SP-A*	*SP-D*
Bacteria			
Escherichia coli J5	Yes	Yes	Yes
Escherichia coli K12	Yes	Yes	?
Group A Streptococcus	Yes	Yes	?
Group B Streptococcus	Yes	Yes	?
Haemophilus influenzae	Yes	Yes	?
Klebsiella species	Yes	Yes	Yes
Listeria monocytogenes	Yes	?	?
Neisseria meningitidis	Yes	?	?
Staphylococcus aureus	Yes	Yes	?
Mycobacteria			
Bacillus Calmette-Guerin	?	Yes	?
Mycobacterium avium	Yes	Yes	Yes
Mycobacterium tuberculosis	?	Yes	Yes
Viruses			
Influenza A	Yes	Yes	Yes
Herpes simplex 2	Yes	Yes	No
HIV-1 and -2	Yes	?	?
Respiratory syncytial virus	?	Yes	Yes
Fungi			
Aspergillus fumigatus	Yes	Yes	Yes
Candida albicans	Yes	No	Yes
Cryptococcus neoformans	Yes	Yes	Yes
Pneumocystis carinii	Yes	Yes	Yes
Saccharomyces cerevisiae	Yes	No	Yes

Details summarized from Clark et al (2004), Hickling et al (2004) and Kishore et al (2006).

modulating immune responses by regulating cytokine production. Alveolar macrophages recruit additional phagocytic cells to the site of pulmonary infections through the release of cytokines (Stockley 1994). The SP-A knockout mouse shows an increase in inflammatory cytokines after challenge by a number of pathogens (Crouch & Wright 2001), indicating the capacity of SP-A to limit the extent of cytokine release.

Excessive local inflammation is a symptom of allergic diseases, such as asthma. SP-A and SP-D also have roles in asthma (Hohlfeld 2002). Evidence now suggests a critical role for SP-A and SP-D in modulation of asthma and allergic disease. The level of SP-A is reduced in patients suffering from asthma. The degree of polymerization of SP-A has been shown to be lower in birch pollen allergy patients than in healthy individuals (Hickling et al 1998). SP-A decreases the proliferative response of dust-mite allergen and PHA-stimulated lymphocytes from children with stable asthma (Wang et al 1998). Allergen-induced bronchial inflammation is associated with a decreased level of SP-A in a murine model of asthma (Hickling et al 2004).

The importance of SP-A and SP-D in allergic processes *in vivo* has been highlighted in mouse models of allergy. Both SP-A and SP-D down-regulate allergic responses to fungal allergens and a recombinant form of SP-D reduced airway hyper-responsiveness in mouse models of allergy to fungal and house dust mite allergens (Hickling et al 2004, Madan et al 2005) SP-A and SP-D may also be involved in antigen presentation via dendritic cells (Brinker et al 2001, 2003). This may provide a mechanism by which they exert anti-allergic effects, for example by favouring polarization of T-helper responses from allergic Th2 responses to more protective Th1 responses (Strong et al 2002). Other possible mechanisms by which SP-A and SP-D exert anti-asthmatic effects may involve direct inhibition of allergen-induced histamine release (Madan et al 2005).

SP-A and SP-D have roles in the clearance of apoptotic cells. This has also been reported for MBL (Vandivier et al 2002). SP-D deficient mice have increased numbers of apoptotic cells in the airways and spontaneously develop emphysema and pulmonary fibrosis (Clark et al 2002), and administration of recombinant SP-D promotes clearance of these cells. Recent reports that collectins bind to nucleic acids may indicate how they interact with apoptotic cells (Palaniyar et al 2004).

Collectin receptors

The interaction between collectins, microorganisms and inflammatory cells is complex and the identification of genuine receptors has been controversial. Several potential receptors for the collectins have been identified, yet strong evidence is available only for one of these candidates. The receptor first described as the collectin receptor in 1990 was shown to be a common receptor for SP-A, MBL, BK and C1q (Malhotra et al 1990). This receptor, which became known as cC1qR ('c' indicating interaction with the collagen region of C1q), was demonstrated to mediate the uptake of opsonized particles into phagocytes. The cC1qR-ligand binding is Ca^{2+} independent and the lectin domain of the collectin is not involved. The receptor cC1qR was subsequently identified as the protein calreticulin (Sim et al 1998). The idea that calreticulin could be a cell surface receptor was initially not widely accepted due to the lack of evidence for a mechanism of cell surface expression for calreticulin. Other candidate receptors emerged, including C1qRp which was reported to be associated with phagocytosis stimulated by C1q, MBL or SP-A. This putative receptor has recently been reported to be an adhesion receptor, identical to CD93, which does not in fact bind directly to any of C1q, MBL or SP-A (McGreal et al 2002). In addition, the complement C3b receptor, CR1, has been reported to interact with C1q and MBL and remains a possible candidate for a universal collectin receptor. Other potential receptors have included a 210 kDa SP-A binding protein from whole rat lung, a 200 kDa SP-A receptor from rat type II cells, a 30 kDa alveolar cell membrane protein identified by anti-idiotype antibodies found in pig and human and gp340, which was found associated with alveolar macrophage and suggested to be an SP-D binding protein (summarised by Hickling et al 2004). None of these potential receptors has been further characterized.

The recent description of a calreticulin/CD91 complex explains how calreticulin may be anchored to the surface of cells. This complex has since been shown to mediate the phagocytic uptake of apoptotic cells via MBL, SP-A, SP-D and C1q (Vandivier et al 2002).

References

Akaiwa M, Yae Y, Sugimoto R et al 1999 Hakata antigen, a new member of the ficolin/opsonin p35 family, is a novel human lectin secreted into bronchus/alveolus and bile. J Histochem Cytochem 47:777–786

Brinker KG, Martin E, Borron P et al 2001 Surfactant protein D enhances bacterial antigen presentation by bone marrow-derived dendritic cells. Am J Physiol Lung Cell Mol Physiol 281:L1453–1463

Brinker KG, Garner H, Wright JR 2003 Surfactant protein A modulates the differentiation of murine bone marrow-derived dendritic cells. Am J Physiol Lung Cell Mol Physiol 284: L232–241

Clark H, Palaniyar N, Strong P, Edmondson J, Hawgood S, Reid KB 2002 Surfactant protein D reduces alveolar macrophage apoptosis in vivo. J Immunol 169:2892–2899

Clark H, Palaniyar N, Hawgood S, Reid KB 2003 A recombinant fragment of human surfactant protein D reduces alveolar macrophage apoptosis and pro-inflammatory cytokines in mice developing pulmonary emphysema. Ann NY Acad Sci 1010:113–116

Clark H, Stehle T, Ezekowitz A, Reid K 2004 Collectins and the acute-phase response. In: Kaufmann SHE, Medzhitov, R, Gordon S (eds) The innate immune response to infection. ASM press, Washington DC, p 199–218

Crouch E, Wright JR 2001 Surfactant proteins a and d and pulmonary host defense. Ann Rev Physiol 63:521–554

Hajela K, Kojima M, Ambrus G et al 2002 The biological functions of MBL-associated serine proteases (MASPs). Immunobiology 205:467–475

Hawgood S, Poulain FR 2001 The pulmonary collectins and surfactant metabolism. Ann Rev Physiol 63:495–519

Hickling TP, Malhotra R, Sim RB 1998 Human lung surfactant protein A (SP-A) exists in several different oligomeric states: oligomer size distribution varies between patient groups. Molec Med 4:265–276

Hickling TP, Clark H, Malhotra R, Sim RB 2004 Collectins and their role in lung immunity. J Leukoc Biol 75:27–33

Hohlfeld JM 2002 The role of surfactant in asthma. Respir Res 3:4

Holmskov U, Malhotra R, Sim RB, Jensenius JC 1994 Collectins: collagenous C-type lectins of the innate immune defence system. Immunol Today 15:67–74

Holmskov U, Thiel S, Jensenius JC 2003 Collections and ficolins: humoral lectins of the innate immune defense. Annu Rev Immunol 21:547–578

Kishore U, Greenhough TJ, Waters P et al 2006 Surfactant proteins SP-A and SP-D: structure, function and receptors. Mol Immunol 43:1293–1315

Krarup A, Thiel S, Hansen A, Fujita T, Jensenius JC 2004 L-ficolin is a pattern recognition molecule specific for acetyl groups. J Biol Chem 279:47513–47519

Larsen F, Madsen HO, Sim RB, Koch C, Garred P 2004 Disease-associated mutations in human mannose-binding lectin compromise oligomerization and activity of the final protein. J Biol Chem 279:21302–21311

Liu Y, Endo Y, Iwaki D et al 2005 Human M-ficolin is a secretory protein that activates the lectin complement pathway. J Immunol 175:3150–3156

Madan T, Kaur S, Saxena S et al 2005 Role of collectins in innate immunity against aspergillosis. Med Mycol 43(Sup1):S155–163

Madsen HO, Satz ML, Hogh B, Svejgaard A, Garred P 1998 Different molecular events result in low protein levels of mannan-binding lectin in populations from southeast Africa and South America. J Immunol 161:3169–3175

Malhotra R, Thiel S, Reid KB, Sim RB 1990 Human leukocyte C1q receptor binds other soluble proteins with collagen domains. J Exp Med 172:955–959

Malhotra R, Haurum J, Thiel S, Sim RB 1992 Interaction of C1q receptor with lung surfactant protein A. Eur J Immunol 22:1437–1445

Malhotra R, Haurum J, Thiel S, Jensenius JC, Sim RB 1993 Pollen grains bind to lung alveolar type II cells (A549) via lung surfactant protein A (SP-A). Biosci Rep 13:79–90

Mayilyan KR, Presanis JS, Arnold JN, Hajela K, Sim RB 2006 Heterogeneity of MBL-MASP complexes. Mol Immunol 43:1286–1292

McGreal EP, Ikewaki N, Akatsu H, Morgan BP, Gasque P 2002 Human C1qRp is identical with CD93 and the mNI-11 antigen but does not bind C1q. J Immunol 168:5222–5232

Palaniyar N, Nadesalingam J, Clark H, Shih MJ, Dodds AW, Reid KB 2004 Nucleic acid is a novel ligand for innate, immune pattern recognition collectins surfactant proteins A and D and mannose-binding lectin. J Biol Chem 279:32728–32736

Petersen SV, Thiel S, Jensenius JC 2001 The mannan-binding lectin pathway of complement activation: biology and disease association. Mol Immunol 38:133–149

Presanis JS, Kojima M, Sim RB 2003 Biochemistry and genetics of mannan-binding lectin (MBL). Biochem Soc Trans 2003 31:748–752

Sim RB, Moestrup SK, Stuart GR et al 1998 Interaction of C1q and the collectins with the potential receptors calreticulin (cC1qR/collectin receptor) and megalin. Immunobiology 199:208–224

Stockley RA 1994 The role of proteinases in the pathogenesis of chronic bronchitis. Am J Respir Crit Care Med 150:S109–113

Strong P, Reid KB, Clark H 2002 Intranasal delivery of a truncated recombinant human SP-D is effective at down regulating allergic hypersensitivity in mice sensitised to allergens of Aspergillus fumigatus. Clin Exp Immunol 130:19–24

Super M, Thiel S, Lu J, Levinsky RJ, Turner MW 1989 Association of low levels of mannan-binding protein with a common defect of opsonisation. Lancet 2:1236–1239

Tang YT, Hu T, Arterburn M et al 2005 The complete complement of C1q-domain-containing proteins in Homo sapiens. Genomics 86:100–111
Teillet F, Dublet B, Andrieu JP, Gaboriaud C, Arlaud GJ, Thielens NM 2005 The two major oligomeric forms of human mannan-binding lectin: chemical characterization, carbohydrate-binding properties, and interaction with MBL-associated serine proteases. J Immunol 174:2870–2877
Turner MW, Hamvas RM 2000 Mannose-binding lectin: structure, function, genetics and disease associations. Rev Immunogenet 2:305–322
Vandivier RW, Ogden CA, Fadok VA et al 2002 Role of surfactant proteins A, D, and C1q in the clearance of apoptotic cells in vivo and in vitro: calreticulin and CD91 as a common collectin receptor complex. J Immunol 169:3978–3986
Wang JY, Shieh CC, You PF, Lei HY, Reid KBM 1998 Inhibitory effect of pulmonary surfactant proteins A and D on allergen-induced lymphocyte proliferation and histamine release in children with asthma. Am J Respir Crit Care Med 158:510–518
Weis WI, Drickamer K 1994 Trimeric structure of a C-type mannose-binding protein. Structure 2:1227–1240

DISCUSSION

Mantovani: Do you get SP-A and SP-D in blood?

R Sim: Uffe Holmskov's group, and others (Sorensen et al 2006, Madsen et al 2003, Eisner et al 2003) have done a lot of measurements of SP-A and SP-D in various body fluids. There is a low level in blood, in the range of a microgram per litre. There is increased blood SP-D in association with some lung inflammation conditions.

Lambrecht: There are several Japanese scientists who use SP-D levels as a marker of disease activity in UIP (usual interstitial pneumonia), a form of lung fibrosis. Is there any link between SP-D and remodelling or restructuring of the alveolar wall after damage? It could be a lattice for Type 2 cells to restore the anatomy of the alveolus after damage.

R Sim: There could be such a link. A lot of work is being done with SP-D in terms of direct effects on cells. The major obstacle in this work is that the full-size recombinant form is not widely available, so that people who are doing experiments can be sure that they have a relevant result with a protein that resembles the native form.

Lambrecht: Does it form lattices in *vivo*?

R Sim: It does have some multimeric but soluble forms. The role of multimerization has not been established. It has been observed in lung lavage fluid that you can get SP-D as a huge oligomer—a fuzzy ball structure that must have 30–40 polypeptide chains in it. But it doesn't appear to form lattice-like oligomers. The fuzzy ball structure will have a radius of something like 90 nm.

Bateman: Do these collectins come from plasma into the alveolar space or from upper airway secretions?

R Sim: SP-A and SP-D are made in alveolar Type 2 cells. Both are also made in other epithelia in small amounts in other parts of the body. In contrast, for mannose-binding lectin (MBL) the major established source of synthesis is the liver.

Bateman: If we try to link SP-A/SP-D with the mucus story we heard earlier, what is the spatial relationship of its airway function? Does it form part of the mucus layer? Does it need mucus for full function?

R Sim: That's an interesting question. It's difficult to envisage in what condition SP-D is within the lung. When we are looking at SP-D, we are looking at it in diluted bronchiolar lavage fluid. The majority of SP-A in the lung (80–90%) is in the tubular myelin layer. It is intercalated in the phospholipids. In bronchoalveolar lavage fluid there is only about 10–20% of the SP-A in the aqueous layer. The SP-D is all in the aqueous layer. However, in producing the bronchoalveolar lavage fluid we have done at least a 30-fold dilution, so any interaction which there would have been with mucus of glycosaminoglycans (GAGs) will have been diluted out.

Sheehan: It is found in mucus, so it is an interesting question to know what form it has in these more restricted, concentrated environments.

Speert: Do you have any sense of how an opsonized mycobacterium entering a macrophage would be handled? Would this trigger a disruptive interaction, or would it be like complement receptor 3-mediated uptake?

R Sim: I can't comment on this. There are a couple of studies on the interaction of SP-D with mycobacterium tuberculosis (Ferguson et al 1999) and also separate studies of SP-A (Lopez et al 2003).

Hoal: The evidence has been that SP-D actually reduces uptake (Ferguson et al 1999). You mentioned the observation that there is a correlation between tuberculosis (TB) and high MBL. Are you referring to serum studies or the genetic work?

R Sim: As far as I know, most of this has been done by genotyping. Genotyping for MBL is more commonly done than determining the phenotype and actual concentration in the blood. It is relatively difficult to set up a sensitive protein concentration assay whereas many laboratories have set up genotyping assays. Generally, the genotype correlates fairly well with MBL concentration: most studies are at least initially done on genotyping.

Hoal: I was wondering whether there was extra evidence that we didn't know about. We did a genotyping study, and also measured serum levels with an ELISA assay, in TB patients and controls. They correlated well and showed what you say (more TB with high MBL levels) (Hoal-van Helden et al 1999). There has been the odd paper that has shown the opposite, though.

Segal: Are there collectins in the mucus of the gastrointestinal (GI) tract?

R Sim: There has been quite a bit of work on SP-A in the GI tract. Again, this is not specifically associating it with mucus. The difficulty is that most biochemists who work on soluble proteins don't also work on mucins or GAGs.

Romani: SP-A and SP-D are also important in the vagina where their levels may affect local immune homeostasis.

Sheehan: There is quite a bit of work on bacteria. I gave an example earlier of how the bacterium phenotype is sensitive to the mucus environment. For a bacterium to change its phenotype is for it to change its surface proteins, e.g. get rid of filopodia. These may be recognition structures and thus it is interesting to consider how the mucus might be predisposing the bacterial phenotype for recognition by particular scavenging agents.

Schoub: Has there been any work demonstrating the level of SP-As and MBLs in the upper part of the respiratory tract? I'm thinking of the nose and nasopharynx. This is where viruses initially establish infection.

R Sim: For MBL in particular, the concentration in spaces in the upper respiratory tract has been measured by Danish colleagues (Garred et al 1993). I am not certain whether there is detailed information like this on SP-D.

Williams: You said that recombinant SP-D could modulate chemokine generation in the lung. Is this a property of the native molecule as well, and what is the mechanism? Is it modulation of phagocytosis, which is a stimulus for chemokine generation?

R Sim: It is difficult to say: the native protein is hard to get hold of, and the recombinant full-size protein is not widely available. Many studies have been done with the recombinant trimeric truncated form, but there is some hesitation about this. It seems to induce effects that might be expected from the native form. There are two recent reviews which include information on chemokine modulation (Kishore et al 2005, 2006).

Latgé: You mentioned that SP-A and SP-D are binding to microorganisms. But you also have surfactant proteins that are hydrophobic. How does all this work when a microorganism gets into the alveolus?

R Sim: We study these proteins rather artificially. The other two proteins, SP-B and SP-C are small hydrophobic proteins, and we don't know of any association between these and SP-A *in vivo*. Certainly, within the lung, most of the SP-A is intercalated into phospholipid, so that at some point SP-A is not principally a soluble protein, but it must be presenting in some way as a fairly regular array anchored into the phospholipid. This would considerably affect how we would envisage the SP-A interaction, with the trapping of microorganisms onto a phospholipid surface by SP-A.

Ryffel: You mentioned that MBL deficiency in patients is common. Do you have any data on the distribution globally?

R Sim: There are three defective variants called B, C and D. Although the frequency of these varies among populations, the cumulative frequency of B plus C plus D is pretty similar. In most populations there is a 20–25% occurrence of these defective genes.

Ryffel: Is this associated with morbidity?

R Sim: Although MBL genotyping has been done to look for association with a wide range of infectious diseases, and there are many studies that show the deficiency has an effect, this is usually fairly small. The main association of the deficiency is with recurrent infection in infants. The strongest association is with recurrent ear infection in infants below the age of 18 months. After this age, when their antibody repertoire is more fully developed, the deficiency of MBL appears to be less important.

McGreal: Several studies have looked at the importance of MBL in immunocompromised individuals, such as patients with neutropenia following chemotherapy for malignancy. The findings from these studies have been mixed. Initial reports indicated that those neutropenic patients who experienced serious infectious episodes had lower MBL concentrations compared to patients who did not (Peterslund et al 2001). Subsequent studies have failed to reproduce these findings in patients being treated for malignancy (Bergmann et al 2003, Kilpatrick et al 2003). However, Mullighan et al (2002) have demonstrated that both donor and recipient MBL genotype is important in preventing infection in patients undergoing haemopoietic stem cell transplantation.

Ryffel: For cystic fibrosis (CF) you mentioned it is often associated with MBL deficiency. Is that primary or secondary?

R Sim: This study was of a large group of CF individuals, among whom decrease in lung function was measured and compared with MBL genotype, i.e. the occurrence of defective genes B, C, D which would be expected to indicate a low level of MBL.

Steinman: How does the MBL see a dying cell? Is the ligand known?

R Sim: The ligand isn't known. There are several studies on the recognition of apoptotic cells by MBL, SP-A and SP-D, but the ligand isn't known for any of them (Ogden et al 2001, Vandiver et al 2002). Peter Garred and Mohammed Daha have shown that MBL does interact with a range of apoptotic cells produced in cell culture, but they indicate that the contribution of MBL to apoptotic cell clearance may be minor (Roos et al 2004, Nauta et al 2003).

Finn: We have been thinking about the possibility that some of the receptors that are suspected to be receptors for these molecules are actually intimately involved in the antigen processing and presentation pathways. We all carry antibodies to these sugars and we call them natural antibodies. We were under the wrong impression for many years that B cells somehow didn't see these sugars. Now there has been a complete change in thinking of how these cells get initially triggered. Even if they make T cell-independent antibodies, they still don't seem to make antibodies on their own. These sugars are presented to them, and in fact there are data that sugars are presented through dendritic cells (DCs) to B cells. This is in an HLA-restricted fashion, but it is not the kind of restriction we think of in terms of peptides in class I and class II molecules. Instead, it is the whole carbohydrate that is associated with the class I and II molecules which helps the T cell and B cell get a closer look at the antigen. You mentioned calreticulin. We don't know what brings these sugars in during their presentation and what takes them back out. Potentially, some of the molecules you described could bring them in and hand them over. Calreticulin may be a very interesting molecule, and could be one of the ways to hold sugars onto the antigen-presenting cells (APCs) to allow the B cell to see them.

Romani: How sensitive are these molecules to modulation of the system? Are there any known modulators?

R Sim: There are a number of conditions where MBL concentration is slightly elevated, and it is sometimes described as a very mild acute phase protein, but this is controversial. Recent work by a Danish group suggests that an increase in MBL is more likely to be caused by thyroid hormones (Riis et al 2005).

References

Bergmann OJ, Christiansen M, Laursen I et al 2003 Low levels of mannose-binding lectin do not affect occurrence of severe infections or duration of fever in acute myeloid leukaemia during remission induction therapy. Eur J Haematol 70:91–97

Eisner MD, Parsons P, Matthay MA, Ware L, Greene K 2003 Acute respiratory distress syndrome network plasma surfactant protein levels and clinical outcomes in patients with acute lung injury. Thorax 58:983–988

Ferguson JS, Voelker DR, McCormack FX, Schlesinger LS 2003 Surfactant protein D binds to Mycobacterium tuberculosis bacilli and lipoarabinomannan via carbohydrate-lectin interactions resulting in reduced phagocytosis of the bacteria by macrophages. J Immunol 163: 312–321

Garred P, Brygge K, Sorensen CH, Madsen HO, Thiel S, Svejgaard A 1993 Mannan-binding protein levels in plasma and upper-airways secretions and frequency of genotypes in children with recurrence of otitis media. Clin Exp Immunol 94:99–104

Hoal-van Helden EG, Epstein J, Victor TC et al 1999 Mannose-binding protein B allele confers protection against tuberculous meningitis. Pediatr Res 45:459–464

Kilpatrick DC, McLintock LA, Allan EK et al 2003 No strong relationship between mannan binding lectin or plasma ficolins and chemotherapy-related infections. Clin Exp Immunol 134:279–284

Kishore U, Bernal AL, Kamran MF et al 2005 Surfactant proteins SP-A and SP-D in human health and disease. Arch Immunol Ther Exp (Warsz) 53:399–417

Kishore U, Greenhough TJ, Waters P et al 2006 Surfactant proteins SP-A and SP-D: structure, function and receptors. Mol Immunol 43:1293–1315

Mullighan CG, Heatley S, Doherty K et al 2002 Mannose-binding lectin gene polymorphisms are associated with major infection following allogeneic hemopoietic stem cell transplantation. Blood 99:3524–3529

Lopez JP, Clark E, Shepherd VL 2003 Surfactant protein A enhances Mycobacterium avium ingestion but not killing by rat macrophages. J Leukoc Biol 74:523–530

Madsen J, Tornoe I, Nielsen O, Koch C, Steinhilber W, Holmskov U 2003 Expression and localization of lung surfactant protein A in human tissues. Am J Respir Cell Mol Biol 29:591–597

Nauta AJ, Raaschou-Jensen N, Roos A et al 2003 Mannose-binding lectin engagement with late apoptotic and necrotic cells. Eur J Immunol 33:2853–2863

Ogden CA, deCathelineau A, Hoffmann PR et al 2001 C1q and mannose binding lectin engagement of cell surface calreticulin and CD91 initiates macropinocytosis and uptake of apoptotic cells. J Exp Med 194:781–795

Peterslund NA, Koch C, Jensenius JC, Thiel S 2001 Association between deficiency of mannose-binding lectin and severe infections after chemotherapy, Lancet 358:637–638

Riis AL, Hansen TK, Thiel S et al 2005 Thyroid hormone increases mannan-binding lectin levels. Eur J Endocrinol 153:643–649

Roos A, Xu W, Castellano G 2004 A pivotal role for innate immunity in the clearance of apoptotic cells. Eur J Immunol 34:921–929

Sorensen GL, Hjelmborg JB, Kyvik KO et al 2006 Genetic and environmental influences of surfactant protein D serum levels. Am J Physiol Lung Cell Mol Physiol 290:L1010–1017

Vandivier RW, Ogden CA, Fadok VA et al 2002 Role of surfactant proteins A, D, and C1q in the clearance of apoptotic cells in vivo and in vitro: calreticulin and CD91 as a common collectin receptor complex. J Immunol 169:3978–3986

Infections and asthma pathogenesis: a critical role for dendritic cells?

Bart N. Lambrecht and Leonie S. van Rijt

Department of Pulmonary Medicine, Erasmus University Medical Center, 3015 GE Rotterdam, The Netherlands

Abstract. Respiratory viral infections can influence the course of asthma at different time points. Severe respiratory viral infections at early age might be associated with a higher prevalence of asthma in later childhood. In established asthma, viral infections are a frequent cause of asthma exacerbation. Epidemiological and experimental animal data can illuminate the mechanisms by which viral infections can lead to sensitization to antigen and exacerbate ongoing allergic airway inflammation. In experimental rodent models of asthma, respiratory viral infection at the time of a first inhaled antigen exposure is described to induce Th2 sensitization and to enhance the allergic response to a second encounter with the same antigen. Virus infections can modulate airway dendritic cell function by up-regulation of costimulatory molecule expression, enhanced recruitment, and by inducing an inflammatory environment, all leading to an enhanced antigen presentation and possibly changing the normal tolerogenic response to inhaled antigen into an immunogenic response. In established asthma, respiratory viral infections attract several inflammatory cells, alter receptor expression on airway smooth muscle and modulate neuroimmune mechanisms, possibly leading to exacerbation of disease. Animal data suggest that the link between respiratory viral infections and increased asthma is causally related, the viral infection acting on the immune and structural cells to enhance antigen presentation and inflammatory cell recruitment.

2006 Innate immunity to pulmonary infection. Wiley, Chichester (Novartis Foundation Symposium 279) p 187–205

Allergic asthma is a chronic inflammatory lung disease characterized by eosinophilic airway inflammation, mucus hypersecretion and airway hyperreactivity in response to inhaled allergens, causing narrowing of the airways. In western societies, there has been a steady increase in the incidence of allergic asthma as well as other allergic diseases such as atopic dermatitis and allergic rhinitis.

Infections can influence the course of asthma at different time points. Severe respiratory viral infections during early age are associated with a higher prevalence of asthma in later childhood, whereas systemic infections caused by mycobacteria or by helminths confer a protective effect on development of allergies and asthma. In established asthma, viral infections are a frequent cause of asthma exacerbation.

Here, we focus on epidemiological and experimental animal data that can illuminate the mechanisms by which infections can enhance or diminish sensitization to antigen, and exacerbate ongoing allergic airway inflammation and focuses on the role played by dendritic cells (DCs).

Influence of infections on development of asthma: epidemiological association studies

Viral respiratory tract infections by influenza virus (INF), respiratory syncitial virus (RSV) and rhinovirus (RV) have been implicated in causing allergic sensitisation and the development of asthma, although their exact role remains controversial. Despite the fact that viral infections influence the immune system in completely different ways at different time points in the asthma pathogenesis, it is clear that they can have great influence on the course of asthma. The role of viral respiratory tract infections in the induction of asthma has been suggested often, although a causal relation has never been documented. Epidemiological studies have shown that severe viral lower respiratory tract infection during the first years of life are associated with an increased risk of developing asthma in children (Illi et al 2001, Sigurs et al 2005). Although lower respiratory tract infection and bronchiolitis caused by RSV poses a risk factor for subsequent wheezing and impaired lung function in early childhood, in most children postbronchiolitic wheezing resolves by 13 years of age (Martinez 2003). 87% of children who had active asthma between the age of 5 and 11 years, had lower respiratory symptoms before the age of 5. In RSV, the association between infection and increased risk for developing asthma holds particularly true for severe infections requiring hospitalization (Sigurs et al 2005). Similarly to RSV infection, the recently described respiratory virus Human Metapneumovirus (hMPV), can cause bronchiolitis and acute wheezing in childhood (Jartti et al 2002, IJpma et al 2004). Further research will have to elucidate whether severe hMPV infection and bronchiolitis at young age are also a risk factor for developing asthma at a later age.

Contrary to the possible enhancing effects of respiratory infections on increased risk of becoming sensitized to aeroallergens, other studies suggest that systemic infections such as tuberculosis, helminth infections or hepatitis A virus infection might offer protection against becoming sensitized. This has been very nicely shown by epidemiological studies showing an inverse risk of allergy with increased tuberculosis notification rates in different countries.

An important question that remains to be answered is the mechanism by which infections in children could be able to facilitate the development of asthma. Th2 sensitization to aeroallergens is a very important risk factor for the development of asthma, and RSV infection has been implicated to enhance Th2 sensitization to aeroallergens (Sigurs 2001). In asthma, unwanted Th2 responses are provoked through exposure to harmless inhaled antigens while in healthy subjects an encounter with the same antigens would lead to immunological tolerance, mediated by regulatory T cells (T_{reg}) (Kuipers & Lambrecht 2004). This would imply that very early in life, viral infections can promote unwanted Th2 immune responses to harmless allergens and influence the process of inhaled tolerance, whereas infections with *Mycobacterium tuberculosis* or *Schistosoma mansonii* might lead to enhanced T_{reg} cell activity or decreased Th2 responses.

Sensitization to inhaled allergen: a process driven by dendritic cells

Most allergic asthma patients have serum IgE specific for several common aeroallergens such as house dust mite, cockroach and animal dander. DCs are essential for priming towards aeroallergens. Upon antigen encounter, they take up, process and transport the allergen to the draining lymph nodes of the lung where it is presented to naïve $CD4^+$ T cells, see Figure 1 (Banchereau et al 2000, Lambrecht 2001, Lambrecht & Hammad 2003, Kuipers & Lambrecht 2004). DCs can regulate the differentiation of T cells into either unresponsive T cells or responsive (Th1/Th2) effector cells. In normal conditions, DCs reside in the airway mucosa and interstitium in an immature state, specialised in taking up antigen across the epithelial barrier, but not yet able to activate naïve T cells, because they lack sufficient expression of costimulatory molecules. It has been stated that DCs need a 'danger signal' to activate T cells sufficiently and thus avoid tolerance (Eisenbarth et al 2002, Kuipers et al 2004). Danger signals are provided by inhaled pathogens, like viruses, bacteria or fungi or derived indirectly from pathogens (LPS, peptidoglycan, etc). Activated effector T cells migrate back to the site of antigen entrance and in case of a second encounter with the antigen, local DCs will stimulate the effector cells to secrete their cytokines. Th2 cells mainly produce interleukin (IL)4, IL5 and IL13, and are crucial in controlling allergic inflammation.

As direct proof that DCs induce Th2 sensitization, we have developed an asthma model in which allergen-pulsed DCs are used to sensitize animals, (Lambrecht et al 2000). Sensitization by DCs induces all the characteristic features of asthma such as airway eosinophilia, Th2 cytokine production by the draining lymph nodes, goblet cell hyperplasia and airway hyperreactivity in response to inhaled allergen (Lambrecht et al 1998, 2000, van Rijt et al 2002, 2005).

Mechanisms by which viral infections enhance DC-driven Th2 sensitization

As DCs are the most important antigen presenting cells in the airways (Holt et al 1988), and these direct the immune response to both viruses and inhaled allergen,

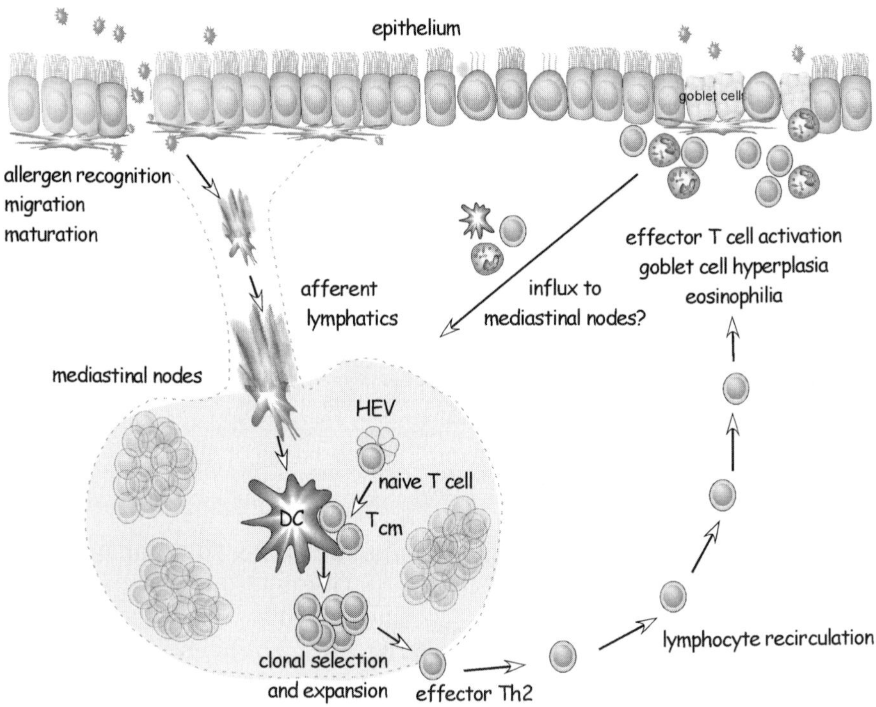

FIG. 1. Role of DCs in asthmatic inflammation. DCs reside in the periphery of the lung, in the airway mucosa and interstitium. Upon recognition of antigen, these cells migrate to the draining lymph nodes where they interact with naïve T cells, selecting out the antigen-specific T cells and inducing their proliferation and differentiation into effector T cells. Effector T cells are mainly generated in response to allergen recognition when the allergen antigen has some component of 'danger' to it, thus inducing the induction of DC maturation. This danger signal can be provided by respiratory viral infections occurring simultaneously with allergen inhalation, or by low grade endotoxin contamination in many allergens. Under particular conditions, allergen presentation by DCs leads to generation of primed Th2 cells that recirculate throughout the body until they encounter allergens again in the periphery. It is now clear that DCs also present allergen to these primed Th2 cells, thus inducing their final effector function. Effector Th2 cells then control the allergic inflammatory reaction leading to goblet cell hyperplasia, eosinophilic influx and bronchial hyperreactivity.

it is intuitive to look more closely at the role that these cells can play in the synergy between respiratory viral infection and allergic sensitisation.

In mice, immunological tolerance is the usual outcome when harmless antigen is inhaled for the first time via the nose. This process is controlled by DCs which produce IL10, and in turn induce IL10 producing regulatory T cells, with a potential to suppress allergic inflammation (Akbari et al 2002). When mice were infected with influenza A, and antigen was administered simultaneously by the nose, inhalational tolerance was broken and instead a strong Th2 response developed (Tsitoura et al 2000, Yamamoto et al 2000). The most likely explanation for these findings is that influenza infection strongly enhanced the function of endogenous lung DCs, as recently shown by Brimnes et al (2003). DCs up-regulate CD40, CD80, CD86 and MHC class II expression after Influenza infection. These mature DCs were shown to be more effective in stimulating naïve antigen specific T cells both *in vivo* and *in vitro* (Brimnes et al 2003). Dendritic cells in interferon (IFN)$\gamma^{-/-}$ mice showed no up-regulation, suggesting the involvement of IFNγ (Dahl et al 2004).

In experimental models, RSV infection can also enhance allergic sensitization. In mice, RSV infection prior to the sensitization protocol increased airway hyperresponsiveness (AHR) and the allergic inflammatory response in the lung (Schwarze et al 1997). In guinea pigs, administration of OVA aerosols induces IgG1 production, AHR and airway inflammation. Simultaneous RSV infection and sensitisation increased IgG1 but did not increase airway inflammation or AHR (Dakhama et al 1999). However, RSV infection during sensitization in mice was also described to decrease the allergic inflammatory response (Peebles et al 2001a, 2001b). A possible explanation for this inconsistency could be the route of sensitization, being either systemic or via the airways. In contrast to several papers dealing with influenza virus, it has not been described what effects RSV infection has on lung DCs. In preliminary experiments, we have demonstrated that RSV can lead to maturation induction in myeloid DC (I. Bogaart, unpublished). *In vitro*, it has also been reported that DCs acquire an immunosuppressive phenotype after infection with RSV and measles (Vidalain et al 2001, Bartz et al 2003).

Induction of maturation in DCs is not the only explanation as to why respiratory viral infections lead to enhanced sensitization. Another subset of DCs, plasmacytoid DCs (pDCs) could divert the tolerogenic response into an active immune response. pDCs are also known as natural IFNα-producing cells, producing copious amounts of this cytokine following viral infection or CpG motif exposure. In the absence of infection, pDCs normally induce tolerogenic immune responses. Recently, we found that pDCs are able to completely prevent eosinophilic airway inflammation after allergen challenge in sensitized mice. Similar to myeloid DCs, pDCs in the lung are able to take up harmless antigen, migrate to the draining lymph nodes to induce tolerance through the induction of regulatory T cells (De Heer et al 2004). The tolerogenic properties of pDCs have been described for a long time. One study has suggested that human pDCs can induce $CD4^+CD25^+$ T regulatory cells (Moseman et al 2004). In the mouse, freshly isolated antigen-pulsed spleen pDCs induce minimal proliferation and no cytokine polarization in antigen-specific T cell receptor transgenic T cells (Martin et al 2002, Boonstra et al 2003). However, activated pDCs can augment cell surface expression of MHC class II and costimulatory molecules, increasing their T cell stimulatory ability and become immunogenic. When pDCs are exposed *in vivo* to bacterial CpG motifs, to influenza or to cytomegalovirus, they become capable of priming $CD8^+$ T cells and induce efficient cytotoxic responses (Cella et al 2000, Dalod et al 2003, Salio et al 2004). From these data, it seems that activated pDCs can present antigens and induce considerable expansion of T cell populations, although less efficiently than myeloid DCs. One mechanism recently raised and that could explain why pDCs can prime T cells is their capacity to differentiate into myeloid DCs upon viral stimulation (Zuniga et al 2004). These data were observed only when very immature bone marrow-derived pDCs were used. More differentiated splenic pDCs were not able to redifferentiate into pDCs. However, whether this really happens in peripheral organs such as the lung remains to be addressed. It is tempting to speculate that this conversion of pDCs of tolerogenic to immunogenic cells could be an explanation as to why infections with respiratory viruses can lead to a break in inhalational tolerance and are often associated with an enhanced allergic response to harmless antigens (Schwarze et al 1997, Yamamoto et al 2000, Dahl et al 2004).

The results obtained by studies in which mice were sensitized via the airways support the finding in the epidemiological studies that respiratory virus infections at a young age can facilitate the sensitization to antigen and/or the strength of the secondary immune response to the antigen. The role of both plasmacytoid and myeloid DCs in this synergy is a very interesting option that clearly needs further study, particularly in human studies.

Mechanisms by which systemic infections reduce DC-driven Th2 sensitization

There are fewer experimental data available as to how systemic infections with *M. tuberculosis* or *Schistosoma mansonii* might lead to a reduced risk of asthma. These and several other commensal microorganisms and chronically infecting pathogens are associated with the induction of tolerance or immune evasion. Indeed, a number of studies have demonstrated that either the whole microorganism or derived compounds can influence myeloid DC function and promote the development of T_{reg} cells. These findings harbour potentially successful adjuvants to battle asthma. However, few of these microbial compounds have been tested for their ability to ameliorate or prevent disease activity *in vivo*. Furthermore, it is still unclear whether the putative protection offered is allergen-specific or unspecific (i.e. via bystander suppression) and whether it will result in long lasting memory. Recent pilot experiments in our laboratory show strong inhibition of asthma development in mice by intratracheally (i.t.) instilled DCs, that were first primed with microbial compounds (BCG, *Bordetella pertussis* filamentous haemaglutinin [FHA] or Prostaglandin D2 [PGD2], a prostaglandin made by *Schistosoma mansoni* [unpublished observations]). *Schistosoma* can produce large amounts of PGD2, slowing down migration and antigen presentation of Langerhans cells. In *Schistosoma* infected individuals, high serum levels of IL10 have been consistently found, together with reduced T cell recall responses to environmental antigens (van den Biggelaar et al 2000). The precise source of IL10 in these patients has not been determined but could be T_{regs}. In house dust mite (HDM)-sensitized children infected with *Schistosoma mansoni*, the wheal and flare reaction and delayed hypersensitivity upon intradermal skin challenge with HDM extract does not occur, despite the presence of high serum levels of HDM-specific IgE (van den Biggelaar et al 2000, Medeiros et al 2004). Whether parasite-derived PGD2 (or other derived signature molecules) might be the explanation as to why *Schistosoma*-infected individuals have a reduced incidence and severity of asthma remains to be addressed (Van Der Kleij et al 2002).

The role of viral airway infections and exacerbations in asthma: epidemiological associations

Despite the fact that asthma is well controlled by steroid treatment, exacerbations of asthma are still very common. Exacerbations often lead to hospital admittance and can result even in death. Although high exposure to allergen in sensitized asthmatics can provoke an exacerbation, virus infections are also major triggers of asthma exacerbations. In 80% of 9–11 year old children (Johnston et al 1995) suffering from an asthma exacerbation and close to half of such episodes in adults (Nicholson et al 1993), viruses could be detected. In children and adults the most frequently identified viruses are rhinovirus and influenza, while in early childhood RSV and parainfluenza are most commonly found. The recently discovered metapneumovirus has also been reported to cause asthma exacerbations in children and adults (Jartti et al 2002).

Mechanisms by which viral infection leads to exacerbations in human asthmatics

A causal relationship between respiratory infection and asthma exacerbations has been studied in human studies of controlled viral infection in mild asthmatics. Experimental inoculations with rhinovirus, a viral pathogen causing the common cold, can induce an increased airway responsiveness in mild asthmatics during acute inflammation which returned to baseline one week after inoculation (Cheung et al 1995) and decreases FEV_1 (forced expiratory volume in one second) with a minimum at 2 days after the inoculation (Grunberg et al 1999). Although individuals with asthma do not have an enhanced susceptibility to rhinovirus infection, they do suffer more from more lower respiratory tract (LRT) infections and have more severe and longer-lasting LRT symptoms (Corne et al 2002). Patients with an asthma exacerbation in which a virus is detected have a higher hospital admission rate and a longer hospital stay. Virus infection and acute asthma is associated with neutrophilic inflammation, cell lysis and more severe clinical disease (Wark et al 2002). A possible explanation could be the increased ICAM1 expression, the main receptor for rhinoviruses, on lower airway epithelium in asthmatics (Manolitsas et al 1994). However, this could not be confirmed in a longitudinal cohort study (Corne et al 2002). Recently, it was shown *in vitro* that epithelial cells of asthmatics undergo less early apoptosis after rhinovirus infection, thereby providing a longer period for viruses to replicate and there was an enhanced virus release in the supernatant. However, this results eventually in a progressive cell lysis (Wark et al 2005). This aberrant behaviour of epithelial cells in asthmatics can contribute to the enhanced virus-induced epithelial damage.

Several human studies indicate that there is a synergistic effect between virus infection and allergen exposure (reviewed in Contoli et al 2005). In the following sections experimental evidence for the possible mechanisms for this synergism are addressed.

Experimental viral airway infection can enhance existing allergic airway inflammation in mice

Results of studies investigating the influence of RSV infection on the strength of an ongoing allergic airway inflammation are more consistent than the results of studies on the link between RSV and sensitization. Allergic sensitization before RSV infection increased AHR, increased proportion of Th2 cytokine producing T lymphocytes during acute infection and increased mucus production in the recovery phase (Robinson et al 1997, Peebles et al 1999, 2001a, Barends et al 2003, 2004, Makela et al 2003, Hashimoto et al 2004). These results suggest that RSV infection can enhance the allergic airway inflammation. Viral infections can enhance the allergic airway inflammation in many ways (Fig. 2).

First, viral exposure will attract several cell types, such as Th1 cells, $CD8^+$ cells and neutrophils, which can intervene with the allergic response. RSV infection induces a Th1 response. Although it was thought that Th1 cells could counterbalance Th2 cells in allergic responses, this has been proven to be incorrect experimentally. The presence of activated Th1 cells during a Th2 mediated airway inflammation did not attenuate the inflammation but caused severe inflammation instead (Hansen et al 1999, Randolph et al 1999). Barends et al (2003) showed that RSV induced IFNγ decreased the allergic disease in mice but IFNγ is not sufficient to prevent allergic airway disease. Influx of RSV-specific Th1 cells could therefore have a deteriorating effect on the inflammation.

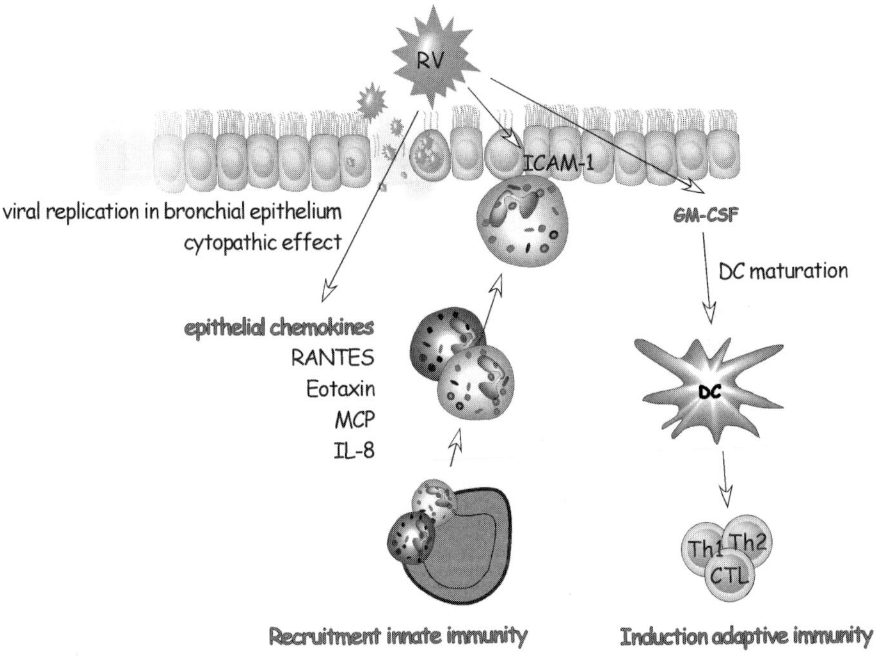

FIG. 2. Exacerbation of existing allergic airway inflammation by (experimental) rhinovirus infection. Rhinovirus attaches to the airway epithelial cell by ICAM1 or the LDL receptor. Infection with rhinovirus increases the expression of ICAM1 leading to an increased susceptibility of the epithelial cells. Replication of rhinovirus in airway epithelium induces the production of several chemokines, which attract cells of the innate and adaptive immune system. The rhinovirus-induced recruitment of eosinophils, mast cells, neutrophils, immature DCs and Th2 cells deteriorates an existing eosinophilic airway inflammation. In addition, infected airway epithelium also secretes GM-CSF, a known maturation factor for DCs, and thereby boosts the adaptive immune response. Matured DCs can instruct new naïve T cells to differentiate into effector T cells.

$CD8^+$ cells produce IFNγ in response to viral exposure to activate macrophages to clear infected cells. However in the presence of IL4, as is the case in the asthmatic lung, $CD8^+$ cells can switch from IFNγ to IL5. IL5 can recruit, activate and enhance survival of eosinophils and in this way enhance the allergic inflammation. Indeed when these $CD8^+$ cells were challenged via the airways with virus peptide, eosinophils were recruited to the airways (Coyle et al 1995).

The immune response to RSV will also attract many neutrophils. The combination of allergen-induced eosinophilia and the recruitment of neutrophils can worsen the airway inflammation, through the action of neutrophilic elastases which are able to degranulate eosinophils (Liu et al 1999). In addition, eosinophils could be involved as an antigen presenting cell in the rhinovirus infection and activating T cells (Handzel et al 1998).

A respiratory tract infection can also attract dendritic cells to the site of inflammation (Gill et al 2005). Mice that were inoculated with influenza virus during the allergic response to OVA developed an increased allergic airway inflammation. DCs acquired a more mature phenotype and induced an enhanced Th1/Th2 response (Dahl et al 2004). During acute infection an enhanced antigen delivery to the draining lymph nodes was observed. This increased migration to the draining lymph nodes of DCs during influenza infection not only increases viral antigen presentation, but also boosts responses against any antigen encountered at the site of inflammation including allergens. In addition, virus-induced inflammation in the lung and draining lymph nodes resulted in the non-specific recruitment of circulating allergen-specific effector/memory cells (Marsland et al 2004).

Secondly, besides the recruitment of intervening cell types to the airways, viral infection can also influence the expression of receptors involved in the allergic response. *In vitro* experiments suggest that rhinovirus infection can enhance bronchial hyperreactivity by up-regulation of the low affinity IgE receptor, FcεRII. Airway smooth muscle (ASM) was cultured in the presence of serum of atopic patients or healthy controls and subsequently inoculated with rhinovirus. ASM cultured in the presence of the serum of asthmatics, increased the expression of low affinity IgE receptor but the rhinovirus infection increased it even more. Binding of IgE on ASM via FcεRII can cause contraction, leading to airway narrowing (Hakonarson et al 1999, Grunstein et al 2001).

Thirdly, viral airway infections can contribute to an enhanced airway inflammation through a neuroimmune mechanism. RSV infection can cause short and long term changes in the distribution and reactivity of sensory nerves across the respiratory tract by the RSV-induced release of nerve growth factor. During RSV infection, stimulation of these nerves causes a marked increase in airway vascular permeability over that in pathogen-free rats and results in an increase in overall inflammatory status. Especially in children, this could play a role because of the different neural wiring (Piedimonte 2003).

Conclusion

In this review, we have summarized the evidence that modulation of DC function by respiratory viruses or by systemic infections is the mechanism behind the known epidemiological link between infection risk and incidence of asthma. In the future it will be important to find those compounds within microbes that could be exploited as therapeutics to halt the ever increasing prevalence of asthma.

Acknowledgements

Leonie S. van Rijt is supported by a EUR fellowship grant and by a BSIK VIRGO consortium grant. B. N. Lambrecht is supported by a Dutch Organisation of Scientific Research (NWO) VIDI grant.

References

Akbari O, Freeman GJ, Meyer EH et al 2002 Antigen-specific regulatory T cells develop via the ICOS-ICOS-ligand pathway and inhibit allergen-induced airway hyperreactivity. Nat Med 8:1024–1032

Banchereau J, Briere F, Caux C et al 2000 Immunobiology of dendritic cells. Annu Rev Immunol 18:767–811

Barends M, Boelen A, de Rond L et al 2003 Respiratory syncytial virus enhances respiratory allergy in mice despite the inhibitory effect of virus-induced interferon-gamma. J Med Virol 69:156–162

Barends M, Van Oosten M, De Rond CG et al 2004 Timing of infection and prior immunization with respiratory syncytial virus (RSV) in RSV-enhanced allergic inflammation. J Infect Dis 189:1866–1872

Bartz H, Turkel O, Hoffjan S, Rothoeft T, Gonschorek A, Schauer U 2003 Respiratory syncytial virus decreases the capacity of myeloid dendritic cells to induce interferon-gamma in naive T cells. Immunology 109:49–57

Boonstra A, Asselin-Paturel C, Gilliet M et al 2003 Flexibility of mouse classical and plasmacytoid-derived dendritic cells in directing T helper type 1 and 2 cell development: dependency on antigen dose and differential toll-like receptor ligation. J Exp Med 197:101–109

Brimnes MK, Bonifaz L, Steinman RM, Moran TM 2003 Influenza virus-induced dendritic cell maturation is associated with the induction of strong T cell immunity to a coadministered, normally nonimmunogenic protein. J Exp Med 198:133–144

Cella M, Facchetti F, Lanzavecchia A, Colonna M 2000 Plasmacytoid dendritic cells activated by influenza virus and CD40L drive a potent TH1 polarization. Nat Immunol 1:305–310

Cheung D, Dick EC, Timmers MC, de Klerk EP, Spaan WJ, Sterk PJ 1995 Rhinovirus inhalation causes long-lasting excessive airway narrowing in response to methacholine in asthmatic subjects in vivo. Am J Respir Crit Care Med 152:1490–1496

Contoli M, Caramori G, Mallia P, Johnston S, Papi A 2005 Mechanisms of respiratory virus-induced asthma exacerbations. Clin Exp Allergy 35:137–145

Corne JM, Marshall C, Smith S et al 2002 Frequency, severity, and duration of rhinovirus infections in asthmatic and non-asthmatic individuals: a longitudinal cohort study. Lancet 359:831–834

Coyle AJ, Erard F, Bertrand C, Walti S, Pircher H, Le Gros G 1995 Virus-specific CD8+ cells can switch to interleukin 5 production and induce airway eosinophilia. J Exp Med 181: 1229–1233

Dahl ME, Dabbagh K, Liggitt D, Kim S, Lewis DB 2004 Viral-induced T helper type 1 responses enhance allergic disease by effects on lung dendritic cells. Nat Immunol 5:337–343

Dakhama A, Bramley AM, Chan NG, McKay KO, Schellenberg RR, Hegele RG 1999 Effect of respiratory syncytial virus on subsequent allergic sensitization to ovalbumin in guinea-pigs. Eur Respir J 13:976–982

Dalod M, Hamilton T, Salomon R et al 2003 Dendritic cell responses to early murine cytomegalovirus infection: subset functional specialization and differential regulation by interferon alpha/beta. J Exp Med 197:885–988

De Heer HJ, Hammad H, Soullie T et al 2004 Essential role of lung plasmacytoid dendritic cells in preventing asthmatic reactions to harmless inhaled antigen. J Exp Med 200:89–98

Eisenbarth SC, Piggott DA, Huleatt JW, Visintin I, Herrick CA, Bottomly K 2002 Lipopolysaccharide-enhanced, toll-like receptor 4-dependent T helper cell type 2 responses to inhaled antigen. J Exp Med 196:1645–1651

Barton T, Ghaffar F, Jafri H et al 2005 Mobilization of plasmacytoid and myeloid dendritic cells to mucosal sites in children with respiratory syncytial virus and other viral respiratory infections. J Infect Dis 191:1105–1115

Grunberg K, Timmers MC, de Klerk EP, Dick EC, Sterk PJ 1999 Experimental rhinovirus 16 infection causes variable airway obstruction in subjects with atopic asthma. Am J Respir Crit Care Med 160:1375–1380

Grunstein MM, Hakonarson H, Hodinka RL, Maskeri N, Kim C, Chuang S 2001 Mechanism of cooperative effects of rhinovirus and atopic sensitization on airway responsiveness. Am J Physiol Lung Cell Mol Physiol 280:L229–238

Hakonarson H, Carter C, Kim C, Grunstein MM 1999 Altered expression and action of the low-affinity IgE receptor FcepsilonRII (CD23) in asthmatic airway smooth muscle. J Allergy Clin Immunol 104:575–584

Handzel ZT, Busse WW, Sedgwick JB et al 1998 Eosinophils bind rhinovirus and activate virus-specific T cells. J Immunol 160:1279–1284

Hansen G, Berry G, DeKruyff RH, Umetsu DT 1999 Allergen-specific Th1 cells fail to counterbalance Th2 cell-induced airway hyperreactivity but cause severe airway inflammation. J Clin Invest 103:175–183

Hashimoto K, Graham BS, Ho SB et al 2004 Respiratory syncytial virus in allergic lung inflammation increases Muc5ac and gob-5. Am J Respir Crit Care Med 170:306–312

Holt PG, Schon-Hegrad MA, Oliver J 1988 MHC class II antigen-bearing dendritic cells in pulmonary tissues of the rat. Regulation of antigen presentation activity by endogenous macrophage populations. J Exp Med 167:262–274

Illi S, von Mutius E, Lau S et al 2001 Early childhood infectious diseases and the development of asthma up to school age: a birth cohort study. BMJ 322:390–395

Ijpma FF, Beekhuis D, Cotton MF et al 2004 Human metapneumovirus infection in hospital referred South African children. J Med Virol 73:486–493

Jartti T, van den Hoogen B, Garofalo RP, Osterhaus AD, Ruuskanen O 2002 Metapneumovirus and acute wheezing in children. Lancet 360:1393–1394

Johnston SL, Pattemore PK, Sanderson G et al 1995 Community study of role of viral infections in exacerbations of asthma in 9–11 year old children. BMJ 310:1225–1229

Kuipers H, Lambrecht BN 2004 The interplay of dendritic cells, Th2 cells and regulatory T cells in asthma. Curr Opin Immunol 16:702–708

Kuipers H, Heirman C, Hijdra D et al 2004 Dendritic cells retrovirally overexpressing IL-12 induce strong Th1 responses to inhaled antigen in the lung but fail to revert established Th2 sensitization. J Leukoc Biol 76:1028–1038

Lambrecht BN 2001 Allergen uptake and presentation by dendritic cells. Curr Opin Allergy Clin Immunol 1:51–59

Lambrecht BN, Hammad H 2003 Taking our breath away: dendritic cells in the pathogenesis of asthma. Nat Rev Immunol 3:994–1003

Lambrecht BN, Salomon B, Klatzmann D, Pauwels RA 1998 Dendritic cells are required for the development of chronic eosinophilic airway inflammation in response to inhaled antigen in sensitized mice. J Immunol 160:4090–4097

Lambrecht BN, De Veerman M, Coyle AJ, Gutierrez-Ramos JC, Thielemans K, Pauwels RA 2000 Myeloid dendritic cells induce Th2 responses to inhaled antigen, leading to eosinophilic airway inflammation. J Clin Invest 106:551–559

Liu H, Lazarus SC, Caughey GH, Fahy JV 1999 Neutrophil elastase and elastase-rich cystic fibrosis sputum degranulate human eosinophils in vitro. Am J Physiol 276:L28–34
Makela MJ, Tripp R, Dakhama A et al 2003 Prior airway exposure to allergen increases virus-induced airway hyperresponsiveness. J Allergy Clin Immunol 112:861–869
Manolitsas ND, Trigg CJ, McAulay AE et al 1994 The expression of intercellular adhesion molecule-1 and the beta 1-integrins in asthma. Eur Respir J 7:1439–1444
Marsland BJ, Scanga CB, Kopf M, Le Gros G 2004 Allergic airway inflammation is exacerbated during acute influenza infection and correlates with increased allergen presentation and recruitment of allergen-specific T-helper type 2 cells. Clin Exp Allergy 34:1299–1306
Martin P, Del Hoyo GM, Anjuere F et al 2002 Characterization of a new subpopulation of mouse CD8alpha+ B220+ dendritic cells endowed with type 1 interferon production capacity and tolerogenic potential. Blood 100:383–390
Martinez FD 2003 Respiratory syncytial virus bronchiolitis and the pathogenesis of childhood asthma. Pediatr Infect Dis J 22:S76–82
Medeiros M Jr, Almeida MC, Figueiredo JP et al 2004 Low frequency of positive skin tests in asthmatic patients infected with Schistosoma mansoni exposed to high levels of mite allergens. Pediatr Allergy Immunol 15:142–147
Moseman EA, Liang X, Dawson AJ et al 2004 Human plasmacytoid dendritic cells activated by CpG oligodeoxynucleotides induce the generation of CD4+CD25+ regulatory T cells. J Immunol 173:4433–4442
Nicholson KG, Kent J, Ireland DC 1993 Respiratory viruses and exacerbations of asthma in adults. BMJ 307:982–986
Peebles RS Jr, Sheller JR, Johnson JE, Mitchell DB, Graham BS 1999 Respiratory syncytial virus infection prolongs methacholine-induced airway hyperresponsiveness in ovalbumin-sensitized mice. J Med Virol 57:186–192
Peebles RS Jr, Hashimoto K, Collins RD et al 2001a Immune interaction between respiratory syncytial virus infection and allergen sensitization critically depends on timing of challenges. J Infect Dis 184:1374–1379
Peebles RS Jr, Sheller JR, Collins RD et al 2001b Respiratory syncytial virus infection does not increase allergen-induced type 2 cytokine production, yet increases airway hyperresponsiveness in mice. J Med Virol 63:178–188
Piedimonte G 2003 Contribution of neuroimmune mechanisms to airway inflammation and remodeling during and after respiratory syncytial virus infection. Pediatr Infect Dis J 22: S66–74; discussion S74–75
Randolph DA, Stephens R, Carruthers CJ, Chaplin DD 1999 Cooperation between Th1 and Th2 cells in a murine model of eosinophilic airway inflammation. J Clin Invest 104: 1021–1029
Robinson PJ, Hegele RG, Schellenberg RR 1997 Allergic sensitization increases airway reactivity in guinea pigs with respiratory syncytial virus bronchiolitis. J Allergy Clin Immunol 100:492–498
Salio M, Palmowski MJ, Atzberger A, Hermans IF, Cerundolo V 2004 CpG-matured murine plasmacytoid dendritic cells are capable of in vivo priming of functional CD8 T cell responses to endogenous but not exogenous antigens. J Exp Med 199:567–579
Schwarze J, Hamelmann E, Bradley KL, Takeda K, Gelfand EW 1997 Respiratory syncytial virus infection results in airway hyperresponsiveness and enhanced airway sensitization to allergen. J Clin Invest 100:226–233
Sigurs N 2001 Epidemiologic and clinical evidence of a respiratory syncytial virus-reactive airway disease link. Am J Respir Crit Care Med 163:S2–6
Sigurs N, Gustafsson PM, Bjarnason R et al 2005 Severe respiratory syncytial virus bronchiolitis in infancy and asthma and allergy at age 13. Am J Respir Crit Care Med 171:137–141

Tsitoura DC, Kim S, Dabbagh K, Berry G, Lewis DB, Umetsu DT 2000 Respiratory infection with influenza A virus interferes with the induction of tolerance to aeroallergens. J Immunol 165:3484–3491

van den Biggelaar AH, van Ree R, Rodrigues LC et al 2000 Decreased atopy in children infected with Schistosoma haematobium: a role for parasite-induced interleukin-10. Lancet 356:1723–1727

van der Kleij D, Latz E, Brouwers JF et al 2002 A novel host-parasite lipid cross-talk. Schistosomal lyso-phosphatidylserine activates toll-like receptor 2 and affects immune polarization. J Biol Chem 277:48122–48129

van Rijt LS, Prins JB, Leenen PJ et al 2002 Allergen-induced accumulation of airway dendritic cells is supported by an increase in CD31(hi)Ly-6C(neg) bone marrow precursors in a mouse model of asthma. Blood 100:3663–3671

van Rijt LS, Jung S, Kleinjan A et al 2005 *In vivo* depletion of lung CD11c+ dendritic cells during allergen challenge abrogates the characteristic features of asthma. J Exp Med 201:981–991

Vidalain PO, Azocar O, Rabourdin-Combe C, Servet-Delprat C 2001 Measle virus-infected dendritic cells develop immunosuppressive and cytotoxic activities. Immunobiology 204:629–638

Wark PA, Johnston SL, Moric I, Simpson JL, Hensley MJ, Gibson PG 2002 Neutrophil degranulation and cell lysis is associated with clinical severity in virus-induced asthma. Eur Respir J 19:68–75

Wark PA, Johnston SL, Bucchieri F et al 2005 Asthmatic bronchial epithelial cells have a deficient innate immune response to infection with rhinovirus. J Exp Med 201:937–947

Yamamoto N, Suzuki S, Shirai A et al 2000 Dendritic cells are associated with augmentation of antigen sensitization by influenza A virus infection in mice. Eur J Immunol 30:316–326

Zuniga EI, McGavern DB, Pruneda-Paz JL, Teng C, Oldstone MB 2004 Bone marrow plasmacytoid dendritic cells can differentiate into myeloid dendritic cells upon virus infection. Nat Immunol 5:1227–1234

DISCUSSION

Hussell: It is amazing how difficult it is to sort out the role of infection in the induction of asthma, especially when this is layered on top of heredity. Can't you take a number of families with a history of asthma and look at the incidence of lower respiratory tract infections in those patients?

Lambrecht: The problem with those sorts of studies is that allergic people are much more sensitive to viral infection. If you give them a common rhinoviral infection which will normally stay only in the upper airways, there has been a recent paper showing that it tends to go deeper (Wark et al 2005). It is hard to look at what is happening. The only way it could be done is experimentally with mouse models of asthma.

Hussell: Could you look at children under one year of age and look at their rate of RSV infection?

Lambrecht: That has been done. It has also been done cross-sectionally. Still, we can't really prove that there is an association. Even in the mice it is controversial. Some people will say that respiratory syncytial virus (RSV) will lead to asthma; others say it is actually protecting.

Bateman: There are two aspects I'd like to comment on. The first is the phasic or temporary nature of each of these influences. Prolonged breast feeding was one of the first factors to be shown to be associated with reduction of risk of atopy. This protection is temporary, and wanes after a few years. In the Lung Health Study 2002, we showed that when parents smoke, the benefit of prolonged breast feeding is abrogated. Thus parental smoking is a confounder in epidemiological studies of risk of allergy. There is also the parasite story. Parasites occur during a particular period of life. Their effect upon atopy might depend on the phase of life at which the parasitism occurs. By the time parasitism is a reality in toddlers, most have already developed atopy and had several viral infection. My second point is that when we studied the ISAAC (International Study of Asthma and Allergies in childhood) data from Cape Town, we went one further step: we divided the children into different bands of socioeconomic deprivation. The rates of asthma and atopic symptoms were much lower in the most socially deprived groups. However, as expected the burden of disease (severity of symptoms) was greater in this group suggesting less access to or utilization of controller therapy. Of great interest to us was that the prevalence of asthma in children who lived in deprived areas but were bussed to good schools was the same as in those from privileged backgrounds, suggesting that this increase in susceptibility to atopic disease occurs rapidly and certainly within one generation.

Lambrecht: One of the nice examples of this is in Switzerland and Austria, where researchers looked at people in the same village, comparing those who live on a farm or not. This will never be picked up with the ISAAC data. Even on such a micro level, children who were raised on a farm—and particularly if their mothers were on the farm during pregnancy—were protected. The effect later waned away. It is very time dependent, and this is something that hasn't been picked up yet with the long cohort studies. Most of these are at age 14. It will be interesting when they are at the irreversible stage of asthma with fixed airway obstruction. All these immunological influences will also reflect on true adult asthma. Most of the childrens' asthma is much more related to allergic asthma. The RSV story might also be a risk factor for adult asthma because it could influence lung growth. This is a messy field.

Didierlaurent: What is the difference between a strong infection and exposure to Toll-like receptor (TLR) agonists or allergens during childhood in terms of asthma incidence?

Lambrecht: The difference is that the epithelium will react totally to a live infection, whereas, for example, the injection of a RSV G protein won't cause the same response as an RSV infection. The epithelium will react very differently. In a real infection secondary factors are induced and there will be shedding of epithelial cells. With innate imprinting you really need a strong stimulus to get that response. This issue might be dependent on having an infection itself. It is going to be harder to find those moieties within those pathogens and then try to mimic the effect.

Didierlaurent: What about epidemiological data?

Lambrecht: The only data that are coming around are on LPS in farms and large communities. In cities in the USA and Canada LPS has been measured: the higher the levels the lower the IgE. LPS and peptidoglycan seem to protect from asthma development.

Speert: I have a question concerning the *Schistosoma mansoni* connection. This is a relatively rare disease, as opposed to *S. haematobium*, which virtually every child gets in areas where this disease is endemic. Might there be something about *S. haemotobium* infection that is not as anti-allergenic?

Lambrecht: I have no idea. There are some data from Southeast Asia where *S. haematobium* is particularly prevalent. The same effect also occurs with *S. haematobium* in that affected individuals have less likelihood of asthma. There are also effects on the age of first infection. If we first induce allergy and then give an infection on top of that, this also dampens the response, because the chronic anti-helminthic T cells have so much IL10 in their response that through bystander suppression asthma symptoms are reduced irrespective of IgE. If you think of it, the children who are in Gabon (where the study was done) are sensitized. They do have house dust mite IgE but this doesn't lead to symptoms or a positive skin test. It could be that infection came later, after the sensitization.

Wilkinson: I have two questions. Mass chemotherapy is an effective way of controlling schistosomiasis, and it is being implemented in many African countries. Do you think this might have unforeseen public health consequences? Second, should small children, who tend not to acquire schistosomiasis until they are large enough to have water contact, be excluded from these programmes so that their risk of atopy decreases?

Lambrecht: There are data from a group that works with Maria Yazdanbakhsh in Brazil. They did a programme in which in certain river areas they gave mass chemotherapy to all the people. In those populations the asthma symptoms start going up. This has been shown also for other helminth infections such as *Ascaris*. In less affluent countries, the burden of disease and severity of asthma is less. The benefit of eradicating the infection at a population level will be much greater than the cost of some asthma cases.

Hussell: In your schistosomiasis allergy model, where there was increased inflammation, the problemme is more one of remodelling and fibrosis rather than inflammation. With the schistosomiasis eggs, there is a big granuloma there that will be causing remodelling and fibrosis. In your dual challenge, did fibrosis go down?

Lambrecht: We haven't looked at that, but I think we need to discriminate what people have done in models where they gave eggs to mice, which is completely different. If you give eggs i.v. you are inducing emboli in the lungs. Some people use this as a Th2/eosinophil model. We are doing this completely differently. Our infection is in the gut and is a live infection. When we did proteomics analysis on bronchoalveolar lavage fluid the proteins we saw in asthmatic mice are exactly the same as those seen with alternative activation of macrophages. We see Ym1, arginaze and Fizz1. These alternatively activated macrophages might form a wall of collagen around large parasites to shield them off. In effect, this occurs also in asthma. The basement membrane is thickening, which could be good for the immune response because the barrier function improves.

Hussell: Have you analysed lung function?

Lambrecht: Yes, it is greatly ameliorated by schistosomiasis. We can also transfer the effect with spleen cells.

Mantovani: What about CR Th2, the DP2 receptor?

Lambrecht: When we gave selective CR Th2 agonist to mice, we didn't suppress the allergy at all. When we give them a secondary response it actually enhances the allergic inflammation. A recent paper by Spik et al (2005) showed that CR Th2 recruits eosinophils. What we do is something rather different. We are trying to elucidate how PGD2 protects by acting on different receptors. This is similar to the recent prostaglandin E2 story where they also show that the EP3 receptor is anti-inflammatory, whereas EP1, 2 and 4 are proinflammatory. A confounding factor is that PGD2 has been claimed to be a bronchoconstrictor in asthmatics mainly through effects on the tromboxane receptor, because it cross reacts with the tromboxane receptor.

Mantovani: There were reports about treating people with helminths: how serious is this?

Lambrecht: People with Crohn's disease have been treated with helminths. From the data we have, this makes sense. But will people be able to control infection? Perhaps in severe patients where there are no other options, we should investigate this.

Steinman: In your system are you treating established asthma, or are you preventing the initial sensitization?

Lambrecht: Particularly in the schisto model, we are preventing sensitization because we also see that IgE levels are completely down from the first ovalbumin boost. However, when we take splenic cells from protected mice and give them after the sensitization in the new host (which has been sensitized to ovalbumin) they suppress inflammation.

Steinman: If the pathway requires IL10, is IL10 needed in both the dendritic cell (DC) and the regulatory cell?

Lambrecht: We don't know.

Walzl: I want to comment on the use of parasites to treat inflammatory bowel disease. The data look quite good. (Summers et al 2003). They use *Trichuris suis*, which is the pig hook worm, which is totally non-pathogenic in humans. This could be developed into a viable therapeutic option.

Lambrecht: We need also to realize that the presence of such a parasite in the gut will also change barrier function dramatically. It could be that the parasite changes the pressure from the colonizing flora. It could be that the parasite is merely changing barrier function.

Segal: We have a study (Marks et al 2006) showing that the major problem in Crohn's disease is a very poor acute inflammatory response. The belief is that as hygiene has increased, the basal state of inflammation in the bowel has fallen. When there is an insult to the bowel, the ability to respond rapidly is compromised. The reason these parasites may work is that they raise the basal level of inflammation to a level such that when other organisms go through the bowel wall, they are more rapidly cleared. I have a question. With regards to the infection and asthma system, is it possible that you have a circulating pool of eosinophils such that if you get a focal infection in the lung, or if you get a parasitic infestation, you then sequester that pool of very reactive eosinophils that are then not available to the airway to respond to allergens? Has anyone looked at the eosinophil pools in the infected and non-infected state?

Lambrecht: Not that I am aware of. If we induce asthma in our mouse models and then look at where the eosinophils come from, most have come from the bone marrow in the last two or three days. Inducing asthma results in an enormous increase of myeloid progenitors in the bone marrow. The bone marrow has so much capacity to generate new eosinophils, that I think they will come fresh from the bone marrow.

Bateman: When one looks at parasitism, levels of total IgE are high and eosinophils are plentiful. These may however have little if anything to do with atopic disease. The idea of eosinophils being sequestered off to other sites of pathology to cause disease is probably not the causal explanation, because it is a very small pool of specifically sensitized eosinophils that are relevant in the pathogenesis of asthma. There are many diseases where there are lots of eosinophils but no atopy. They are important cells when they are present, but they are not essential. There are neutrophilic forms of asthma. The parasite story is good and plausible, but the epidemiological data are mixed on this. Some studies have been negative, showing no relationship between atopy and parasitism. In one South African study sensitization to *Ascaris lumbricoides* antigens was associated with a higher prevalence of asthma, and antigen inhalation provocation with *Ascaris* antigens were positive in some patients with cutaneous sensitization to antigen (Joubert et al 1979).

Lawn: To address the question of whether schistosomiasis has a direct immunological effect or rather disrupts mucosal barrier function, it would be useful to compare *S. mansoni* and *S. haemotobium*, which have similar systemic immunological effects but affect different anatomical sites with *S. mansoni* predominantly affecting the gastrointestinal mucosa.

Romani: Are the regulatory T cells in your model directly originated in the lung or do they migrate from the gut, as has been shown in other models?

Lambrecht: We haven't characterized them yet. They could come from the spleen.

References

Joubert JR, de Klerk HC, Malan C 1979 Ascaris lumbricoides and allergic asthma: A new perspective. S Afr Med J 15:599–602

Marks DJ, Harbord MW, MacAllister R et al 2006 Defective acute inflammation in Crohn's disease: a clinical investigation. Lancet 367:668–678

Spik I, Brenuchon C, Angeli V et al 2005 Activation of the prostaglandin D2 receptor D2 receptor DP2/CRTH2 increases allergic inflammation in mouse. 174:3703–3708

Summers RW, Elliott DE, Qadir K, Urban JF Jr, Thompson R, Weinstock JV 2003 Trichuris suis seems to be safe and possibly effective in the treatment of inflammatory bowel disease. Am J Gastroenterol 98:2034–2041

Trottein F, Mallevaey T, Faveeuw C, Capron M, Leite-de-Moraes M 2006 Role of the natural killer T lymphocytes in Th2 responses during allergic asthma and helminth parasitic diseases. Chem Immunol Allergy. 90:113–127

Wark PA, Johnston SL, Bucchieri F et al 2005 Asthmatic bronchial epithelial cells have a deficient innate immune response to infection with rhinovirus. J Exp Med 201:937–947

Innate and adaptive immunity in lung cancer

L. A. Vella and O. J. Finn

Department of Immunology, E1040 Biomedical Science Tower, University of Pittsburgh, School of Medicine, Pittsburgh, PA 15261, USA

Abstract. The immune system is alerted to the presence of a pathogen through the activation of the innate immune system. The message is transmitted to the cells of the adaptive immunity through activated antigen-presenting cells. The development of specific immunity capable of eliminating the pathogen is orchestrated by cytokines and chemokines produced by the innate system. When everything functions optimally, the pathogen is eradicated and specific memory response is established. This finely tuned system can be subverted by pathogens, leading to disease. Immunity to cancer is orchestrated in the same way and it is now recognized that the early stages of tumour development are recognized by the cells of innate immunity that transmit this message to the cells of adaptive immunity. The molecules that alert the immune system and are also its targets are tumour antigens. Two important antigens for lung tumour-specific immunity are MUC1 and cyclin B1. We discuss how each molecule interacts with the innate and the adaptive immunity and the types of the immune responses that result for these interactions. We also discuss the state of immunosuppression of adaptive immunity in cancer patients due to chronic activation of the innate immune system.

2006 Innate immunity to pulmonary infection. Wiley, Chichester (Novartis Foundation Symposium 279) p 206–215

In industrialized nations, 19% of adult deaths are attributable to lung cancer (Ezzati & Lopez 2003). The high mortality associated with lung cancer is due to the lack of effective screening for early disease and failure of current treatment modalities to cure already advanced disease at diagnosis. Better control of the disease process will come with the development of new tools for early diagnosis and new approaches to therapy and prevention of lung cancer and other cancers (Finn 2005). The immune system can potentially be exploited for both (Finn 2005).

The function of the innate and the adaptive immune systems is intimately involved in cancer development (Dunn et al 2004). Growing evidence supports a detrimental role of chronic inflammation in cancer development. Chronic inflammatory conditions such as inflammatory bowel disease and Hashimoto's thyroiditis are associated with an increased risk of colon cancer and thyroid cancer, respectively (Dailey et al 1955, Dobbins 1984). In animal models of inflammation and cancer, genetic mutations leading to cancer initiation promote an inflammatory response that—when unable to clear the lesion effectively—serves to promote neoplastic transformation (de Visser & Coussens 2005). In the case of lung cancer, the mutagenic and inflammatory effects of smoking set the stage for a similar type of cancer-promoting immune response. The same appears to be the case for some lung pathogens such as *Chlamidia pneumoniae* (Littman et al 2005) or Pneumocystis (de la Horra et al 2004). Relatively little is known so far about the prognostic significance of the human immune response to lung cancer, both innate and adaptive. We will review here the published and ongoing studies on this subject and use the data obtained so far in support of cancer vaccines as a way to tip the balance between innate and adaptive immunity in favour of adaptive immunity and successful cancer elimination.

Innate immunity in lung cancer

The earliest pathological feature of cigarette smoking is an inflammatory infiltrate that extends throughout the lung tissue (Dobbins 1984). Since smoke inhalation repeatedly damages epithelial cells of the lung, effector cells of the innate immune system are constantly recruited and continuously activated, setting the stage for chronic inflammation that never resolves. This is also the ideal setting for promotion of cancers that are initiated by cigarette carcinogens.

Tumour-associated macrophages (TAMs) and mast cells (MCs) are two of the best-documented innate immune facilitators of lung cancer progression. In a study of non-small cell lung cancers (NSCLCs) from 48 patients, 1/3 of the immune cells in the tumour stroma were determined to be TAMs (Kataki et al 2002). While the normal role of macrophages in the acute response is to recruit and stimulate effector cells, the presence of TAMs in many different human tumours has been associated with a worse prognosis for the large majority of cancer types (Bingle et al 2002). TAMs have been repeatedly correlated with increased angiogenesis, invasiveness and metastasis (Li et al 2002, Tsutsui et al 2005, Varney et al 2005). The same is true in NSCLCs, where TAMs have been shown to correlate with worse prognosis and increased angiogenesis (Koukourakis et al 1998, Takanami et al 1999).

The MC is another innate immune effector that may facilitate lung cancer development in the setting of chronic inflammation. Many studies show strong evidence that tumor-infiltrating MCs are negative prognostic factors in cancer patients, including those with breast, cervical, gastric, skin and lung cancers (Ribatti et al 2001). Further causative links between MCs and cancer have been made through mouse models of squamous epithelial cancers (Coussens et al 1999). In lung cancer, the MC infiltrate is elevated, and the numbers of MCs increase as the stage of disease increases (Imada et al 2000, Tataroglu et al 2004). In NSCLC, MC number also correlates with microvessel density, and in lung adenocarcinomas, the presence of MCs correlates with tumour progression and worse survival (Takanami et al 2000).

The only innate immune cell that is correlated with positive prognoses in lung cancer is the natural killer cell (NK). While not antigen-specific responders, NK cells have surface molecules that recognize specific features of preneoplastic or neoplastic cells. Studies in lung squamous cell carcinoma and lung adenocarcinoma have shown that the presence of NK cells in the tumour microenvironment correlates with increased patient survival after surgical tumour resection (Takanami et al 2001, Villegas et al 2002).

In addition to the chronic innate inflammation at the tumour site that can promote tumour growth, we and others have shown that the adaptive immune system in cancer patients is profoundly suppressed (Ochoa & Longo 1995, Schmielau et al 2001, Whiteside 2006) and cannot participate in tumour rejection. This suppression is manifest in the low cytokine production of activated T cells, low antibody production and ineffective isotype switching by B cells, and increased apoptosis of lymphocytes. We discovered that one of the mechanisms that promotes this suppression of the adaptive immunity is the state of systemic chronic activation of granulocytes (Schmielau et al 2001).

Adaptive immunity in lung cancer

The adaptive immune response to lung cancer—composed of B cells, T cells and antibodies—has long been studied for use in cancer prognosis and cure. Unlike the chronic inflammatory (non-specific) immune response, T cell infiltration of tumours has been shown to correlate with enhanced post-surgical survival of lung cancer patients (Johnson et al 2000, Wakabayashi et al 2003). Further, patients with higher numbers of T cells in the lung tumour stroma had a lower size and stage of cancer than did patients with a low T cell infiltrate (Eerola et al 2000). Still, the specificities of the T cell responders were not identified in these studies; therefore, while the studies demonstrate that specific responders are beneficial, they do not identify the antigen against which the specific response is directed. In order to develop more precise immunological/prognostic measurements and to identify the best components of a lung cancer vaccine, it is important to identify the lung tumour antigens that elicit the strongest response, leading to the best prognostic outcomes.

Of the many lung tumour antigens identified, immune responses against very few have been associated with clinical outcome. Moreover, studies that assess the prognostic value of anti-tumour-antigen immune responses in lung cancer have involved the antibody response alone; the prognostic value of known antigen-specific T cell responses in lung cancer has yet to appear in the literature.

One potentially important tumour antigen in lung cancer, cyclin B1(CB1), was initially described by us (Kao et al 2001). CB1 is a cell cyclin required for the transition from G2 to M phase of the cell cycle. In a normal cell, CB1 is expressed very transiently as the cell moves into mitosis, after which it is immediately ubiquitinated for proteasomal degradation (Pines & Hunter 1992). In several cancers, including lung cancer, CB1 is constitutively overexpressed, and it accumulates in the cytoplasm. CB1 overexpression has since been correlated with a poorer patient prognosis in lung, oesophageal and tongue cancers (Hassan et al 2001, Nozoe et al 2002, Soria et al 2000). We have found that CB1 elicits both memory T cell and antibody responses in cancer patients (Kao et al 2001). T cell-dependent antibodies against CB1 have also been detected by us and others in hepatocellular, breast, gastric, prostate, colorectal and lung cancers (Covini et al 1997, Suzuki et al 2005).

Thus far, only three tumour antigens have been shown to link anti-tumour immune responses to clinical significance in lung cancer: p53, MUC1 and the neuronal protein Hu. Mutations in the tumour suppressor gene p53 are common in lung cancer, and those that lead to aberrant expression of the p53 protein can provoke immune responses to this protein. The value of anti-p53 antibodies has been extensively evaluated in NSCLC and in some cases confirmed. Following a cohort of asbestos-exposed individuals, one group found that an increase in anti-p53 antibodies predicted eventual cancer development, with 76% of patients who tested positive for antibody developing NSCLC in an average of 3.5 years (Li et al 2005). In another study, the presence of serum anti-p53 antibody in patients with NSCLC correlated with enhanced response to radiotherapy (Bergqvist et al 1998). However, anti-p53 antibodies were a negative prognostic factor in late stage (III/IV) adenocarcinoma and/or had no correlation in late stage squamous cell lung carcinomas (Bergqvist et al 2004). These mixed results may be due to the fact that p53 mutations come in many different forms. While some lung cancers may have mutations that lead to overexpressed or altered (therefore immunogenic) p53, others have deletion mutations where no p53 antigen exists for immunological recognition (Brambilla & Brambilla 1997). In that respect, it is important to point out that cyclin B1 is overexpressed—and subsequently immunogenic—as a result of all types of p53 deregulation, including deletion (Innocente et al 1999, Yu et al 2002). Therefore, antibodies against CB1 may be a better method of screening and prognosis that would cover a larger number of patients and tumours with p53 mutations.

Another tumour antigen with known clinical significance in NSCLC is MUC1, initially described by us as a target of human immune responses (Barnd et al 1989). A study of 60 NSCLC patients demonstrated that patients considered high responders for anti-MUC1 antibody had a significantly improved survival rate during a 54 month follow-up period (Hirasawa et al 2000). These data parallel the results seen with assessment of antibodies against the neurologic protein Hu. Anti-Hu antibodies were initially discovered as the cause of immune cross-reactivity in patients with small cell lung cancer (SCLC) and the autoimmune paraneoplastic neurological disorders (PND) (Graus et al 1987). Surprisingly, further analysis showed anti-Hu antibodies were mostly present in SCLC patients who did not have PND (Dalmau et al 1990). The presence of anti-Hu antibody correlated with a lower stage of disease, better response to chemotherapy, and better overall survival (Graus et al 1997). However, a more recent study found no correlation between anti-Hu antibody and SCLC prognosis (Monstad et al 2004) suggesting the need for additional evaluation of this particular antigen.

Restoring the balance

Inasmuch as the non-specific activation of the innate immune system leading to chronic inflammation has been shown to facilitate lung cancer progression and to suppress the tumour antigen-specific activation of the adaptive immune system, it may be predicted that restoring the function of the adaptive immunity and establishing the proper balance between the innate and the adaptive systems could be the key to cancer control. Vaccination against one or a panel of tumour antigens, such as those described above, is the best way to induce antigen-specific adaptive immunity and in that way tip the balance towards an effective, anticancer immune response (Atanackovic et al 2004, Chang et al 2005).

References

Atanackovic D, Altorki NK, Stockert E et al 2004 Vaccine-induced CD4+ T cell responses to MAGE-3 protein in lung cancer patients. J Immunol 172:3289–3296

Barnd DL, Lan MS, Metzgar RS, Finn OJ 1989 Specific, major histocompatibility complex-unrestricted recognition of tumor-associated mucins by human cytotoxic T cells. Proc Natl Acad Sci USA 86:7159–63

Bergqvist M, Brattstrom D, Larsson A et al 1998 P53 auto-antibodies in non-small cell lung cancer patients can predict increased life expectancy after radiotherapy. Anticancer Res 18:1999–2002

Bergqvist M, Brattstrom D, Larsson A, Hesselius P, Brodin O, Wagenius G 2004 The role of circulating anti-p53 antibodies in patients with advanced non-small cell lung cancer and their correlation to clinical parameters and survival. BMC Cancer 4:66

Bingle L, Brown NJ, Lewis CE 2002 The role of tumour-associated macrophages in tumour progression: implications for new anticancer therapies. J Pathol 196:254–265

Brambilla E, Brambilla C 1997 p53 and lung cancer. Pathol Biol (Paris) 45:852–863

Chang GC, Lan HC, Juang SH et al 2005 A pilot clinical trial of vaccination with dendritic cells pulsed with autologous tumor cells derived from malignant pleural effusion in patients with late-stage lung carcinoma. Cancer 103:763–771

Coussens LM, Raymond WW, Bergers G et al 1999 Inflammatory mast cells up-regulate angiogenesis during squamous epithelial carcinogenesis. Genes Dev 13:1382–1397

Covini G, Chan EK, Nishioka M, Morshed SA, Reed SI, Tan EM 1997 Immune response to cyclin B1 in hepatocellular carcinoma. Hepatology 25:75–80

Dailey ME, Lindsay S, Skahen R 1955 Relation of thyroid neoplasms to Hashimoto disease of the thyroid gland. AMA Arch Surg 70:291–297

Dalmau J, Furneaux HM, Gralla RJ, Kris MG, Posner JB 1990 Detection of the anti-Hu antibody in the serum of patients with small cell lung cancer-a quantitative Western blot analysis. Ann Neurol 27:544–552

de la Horra C, Varela JM, Fernandez-Alonso J et al 2004 Association between human-Pneumocystis infection and small-cell lung carcinoma. Eur J Clin Invest 34:229–235

de Visser KE, Coussens LM 2005 The interplay between innate and adaptive immunity regulates cancer development. Cancer Immunol Immunother 54:1143–1152

Dobbins WO 3rd 1984 Dysplasia and malignancy in inflammatory bowel disease. Annu Rev Med 35:33–48

Dunn GP, Old LJ, Schreiber RD 2004 The three Es of cancer immunoediting. Annu Rev Immunol 22:329–360

Eerola AK, Soini Y, Paakko P 2000 A high number of tumor-infiltrating lymphocytes are associated with a small tumor size, low tumor stage, and a favorable prognosis in operated small cell lung carcinoma. Clin Cancer Res 6:1875–1881

Ezzati M, Lopez AD 2003 Estimates of global mortality attributable to smoking in 2000. The Lancet 362:847

Finn OJ 2005 Immune response as a biomarker for cancer detection and a lot more. N Engl J Med 353:1288–1290

Graus F, Elkon KB, Lloberes P et al 1987 Neuronal antinuclear antibody (anti-Hu) in paraneoplastic encephalomyelitis simulating acute polyneuritis. Acta Neurol Scand 75:249–252

Graus F, Dalmou J, Rene R et al 1997 Anti-Hu antibodies in patients with small-cell lung cancer: association with complete response to therapy and improved survival. J Clin Oncol 15:2866–2872

Hassan KA, El-Naggar AK, Soria JC, Liu D, Hong WK, Mao L 2001 Clinical significance of cyclin B1 protein expression in squamous cell carcinoma of the tongue. Clin Cancer Res 7:2458–2462

Hirasawa Y, Kohno N, Yokoyama A, Kondo K, Hiwada K, Miyake M 2000 Natural autoantibody to MUC1 is a prognostic indicator for non-small cell lung cancer. Am J Respir Crit Care Med 161:589–594

Imada A, Shijubo N, Kojima H, Abe S 2000 Mast cells correlate with angiogenesis and poor outcome in stage I lung adenocarcinoma. Eur Respir J 15:1087–1093

Innocente SA, Abrahamson JLA, Cogswell JP, Lee JM 1999 PNAS 96:2147–2152

Johnson SK, Kerr KM, Chapman AD et al 2000 Immune cell infiltrates and prognosis in primary carcinoma of the lung. Lung Cancer 27:27–35

Kao H, Marto JA, Hoffmann TK et al 2001 Identification of cyclin B1 as a shared human epithelial tumor-associated antigen recognized by T cells. J Exp Med 194:1313–1323

Kataki A, Scheid P, Piet M et al 2002 Tumor infiltrating lymphocytes and macrophages have a potential dual role in lung cancer by supporting both host-defense and tumor progression. J Lab Clin Med 140:320–328

Koukourakis MI, Giatromanolaki A, Kakolyris S et al 1998 Different patterns of stromal and cancer cell thymidine phosphorylase reactivity in non-small-cell lung cancer: impact on tumour neoangiogenesis and survival. Br J Cancer 77:1696–1703

Li C, Shintani S, Terakado N, Nakashiro K, Hamakawa H 2002 Infiltration of tumor-associated macrophages in human oral squamous cell carcinoma. Oncol Rep 9:1219–1223

Li Y, Karjalainen A, Koskinen H et al 2005 p53 autoantibodies predict subsequent development of cancer. Int J Cancer 114:157–160

Littman AJ, Jackson LA, Vaughan TL 2005 Chlamydia pneumoniae and lung cancer: epidemiologic evidence. Cancer Epidemiol Biomarkers Prev 14:773–778

Monstad SE, Drivsholm L, Storstein A et al 2004 Hu and voltage-gated calcium channel (VGCC) antibodies related to the prognosis of small-cell lung cancer. J Clin Oncol 22:795–800

Nozoe T, Korenaga D, Kabashima A, Ohga T, Saeki H, Sugimachi K 2002 Significance of cyclin B1 expression as an independent prognostic indicator of patients with squamous cell carcinoma of the esophagus. Clin Cancer Res 8: 817–822

Ochoa AC, Longo DL 1995 Alteration of signal transduction in T cells from cancer patients. Important Adv Oncol 43–54

Pines J, Hunter T 1992 Cyclins A and B1 in the human cell cycle. Ciba Found Symp 170:187–196 discussion 196–204

Ribatti D, Vacca A, Nico B, Crivellato E, Roncali L, Dammacco F 2001 The role of mast cells in tumour angiogenesis. Br J Haematol 115:514–521

Schmielau J, Nalesnik MA, Finn OJ 2001 Suppressed T-cell receptor zeta chain expression and cytokine production in pancreatic cancer patients. Clin Cancer Res 7:933s–939

Soria JC, Jang SJ, Khuri FR et al 2000 Overexpression of cyclin B1 in early-stage non-small cell lung cancer and its clinical implication. Cancer Res 60:4000–4004

Suzuki H, Graziano DF, McKolanis J, Finn OJ 2005 T cell-dependent antibody responses against aberrantly expressed cyclin B1 protein in patients with cancer and premalignant disease. Clin Cancer Res 11:1521–1526

Takanami I, Takeuchi K, Kodaira S 1999 Tumor-associated macrophage infiltration in pulmonary adenocarcinoma: association with angiogenesis and poor prognosis. Oncology 57:138–142

Takanami I, Takeuchi K, Naruke M 2000 Mast cell density is associated with angiogenesis and poor prognosis in pulmonary adenocarcinoma. Cancer 88:2686–2692

Takanami I, Takeuchi K, Giga M 2001 The prognostic value of natural killer cell infiltration in resected pulmonary adenocarcinoma. J Thorac Cardiovasc Surg 121:1058–1063

Tataroglu C, Kargi A, Ozkal S, Esrefoglu N, Akkoclu A 2004 Association of macrophages, mast cells and eosinophil leukocytes with angiogenesis and tumor stage in non-small cell lung carcinomas (NSCLC). Lung Cancer 43:47–54

Tsutsui S, Yasuda K, Suzuki K, Tahara K, Higashi H, Era S 2005 Macrophage infiltration and its prognostic implications in breast cancer: the relationship with VEGF expression and microvessel density. Oncol Rep 14:425–431

Varney ML, Johansson SL, Singh RK 2005 Tumour-associated macrophage infiltration, neovascularization and aggressiveness in malignant melanoma: role of monocyte chemotactic protein-1 and vascular endothelial growth factor-A. Melanoma Res 15:417–425

Villegas FR, Coca S, Villarrubia VG et al 2002 Prognostic significance of tumor infiltrating natural killer cells subset CD57 in patients with squamous cell lung cancer. Lung Cancer 35:23–28

Wakabayashi O, Yamazaki K, Oizumi S et al 2003 CD4+ T cells in cancer stroma, not CD8+ T cells in cancer cell nests, are associated with favorable prognosis in human non-small cell lung cancers. Cancer Sci 94:1003–1009

Whiteside TL 2006 Immune suppression in cancer: effects on immune cells, mechanisms and future therapeutic intervention. Semin Cancer Biol 16:3–15

Yu M, Zhan Q, Finn OJ 2002 Immune recognition of cyclin B1 as a tumor antigen is a result of its overexpression in human tumors that is caused by non-functional p53. Mol Immunol 38:981–987

DISCUSSION

Lambrecht: One strategy you didn't mention is that some people think that this therapy might be more successful if you first get rid of all the immunosuppressing factors. One way of doing this would be in an adjuvant setting following curative surgery, which is why your post-pancreatic surgery was so nice. For lung cancer this is possible. But wouldn't it be logical to add vascular endothelial growth factor (VEGF) receptor antagonists, anti-interleukin (IL)6 or COX2 inhibition?

Finn: Most of this work is currently being done in animal models rather than in clinical trials. This is confounded by the fact that some of the inhibitors, if they have any activity on their own, have to be used as a single agent first. For example, a non-steroidal anti-inflammatory drug has just been approved for treatment of pancreatic cancer patients because in randomized trials it prolonged life by 12 days. Now we have to allow this to be administered first before the vaccine, so the patients are even further away from being able to be helped by immune therapy. Not all tumours are equal in terms of inducing regulatory T cells. Ovarian cancer is notorious for inducing T_{reg}, but we don't really find them in pancreatic cancer. From the perspective of a vaccine this is an important question. We now have a T cell receptor transgenic mouse for MUC1. 100% of the T cells in that mouse are specific for the MUC1 peptide in MHC class II. When we load this peptide on dendritic cells (DCs) and mature them, and then vaccinate, in the same mouse you generate effector cells and regulatory cells. Your success depends on the difference between the two responses. The larger the difference, the better response. Every vaccination protocol generates T_{reg}. We have to be smart in vaccinating in a way that will give us a big difference between the effector and regulatory cells.

Steyn: Lynn Wilson and colleagues, using prostatic cancer as a model propose that there are cancer stem cells (Salm et al 2005). This implies that our therapies are aimed at the wrong cells: these are the progeny of the stem cells, and targeting them won't prevent relapses and metastases. Would your approach recognize such stem cells?

Finn: I would have paid you to ask that question!

Steyn: You still can.

Finn: The recent focus on tumour stem cells is a wonderful support for the need to develop immunotherapy or cancer vaccines. Both MUC1 and cyclin B1 are expressed on tumour stem cells. The stem cells aren't proliferating and thus are not targeted by chemotherapy. They aren't doing much with kinases either as they are in a resting state, so they cannot be targeted by protein kinase inhibitors. But, they have these tumour antigens and so they should be picked up by the immune system.

Steinman: Can you tell us more about the immune responses in the patients with polyps? Do you see T cell responses only in the patients who have antibody?

Finn: The polyp study is a retrospective study because we have a huge number of serum samples stored from a wide range of patients. We are looking at the types of antibody as a surrogate marker for T cell activation. MUC1 is very good at stimulating B cells to make IgM because it has a tandem repeat structure. In cancer patients and other immunosuppressed individuals you can generate nice IgM responses to MUC1, and they don't do very much. We have no IgM at all in the polyp patients. We have IgG and IgA, a very different and potentially protective immune response.

Steinman: Have you tried to look for the T cells yet?

Finn: We are beginning a prospective study with 100 patients to look at the T cells directly.

Steinman: Is the isotype of the antibody useful for giving you information about the type of T cell?

Finn: The problem is that while the isotypes are clearly defined in the animal models, in humans it is hard to know which one we really want. We are looking at all the isotypes but for now only for the sake of information, hoping that the right one will fall out. As it falls out, we will know what we want to generate through a vaccine.

Romani: Can you measure the level of IFNγ by ELISA together with isotypes?

Finn: Yes. We did this and it led us to a wrong conclusion in patients. We looked for IFNγ-dependent isotypes, thinking that this will indicate activation of the Th1 response. But in cancer patients there is tremendous activation of NK cells. They are pumping out tons of IFNγ and they are helping the B cells switch to IgG3, for example, in the animal models. What happens in this case is that there is no tumour rejection response. These tumours are susceptible to T cells but not really antibodies. For a while we thought we had it all, and that it was a superactivation of NK cells, but this was not therapeutic. We are looking but we can't really draw the conclusion until we show that it is a T cell that is producing IFNγ.

Mantovani: Coming back to the connection with inflammation, I guess that the neutrophils have degranulated. Did you check for elastase in the serum of the patients? This would be a direct marker.

Finn: We didn't. We used isoprostane because it is the longest-lasting marker. We have serum from patients at different times post surgery. We can recapitulate everything *in vitro* just by taking healthy donor's blood and then exposing T cells to the products of the activated PMNs. We get exactly the same activation-induced cell death of T cells and suppression of cytokine production.

Steinman: You have T cell receptor transgenics to MUC1. If you put these T cells into a mouse, is the MUC1 being presented in the steady state?

Finn: No. We don't find any in the normal colon.

Lambrecht: What happens during radiation-induced damage in those mice? Is it presented then?

Finn: The only time that MUC1 is a target of an immune response, other than cancer, is on the lactating epithelium. Hyperproliferation of epithelial cells induces a high level of MUC1. It was originally discovered in the milk fat globules. It has a lot of unglycosylated sites, and there has been a lot of connection between reduction of cancers and multiple lactations. We did a large case- control study where there were several non- cancerous conditions where MUC1 could have been presented to the immune system, and it could have been presented in an inflammatory state. We found tremendous correlation with reduced risk for ovarian cancer (Cramer et al 2005).

References

Cramer DW, Titus-Ernstoff L, McKolanis JR et al 2005 Conditions associated with antibodies against the tumor-associated antigen MUC1 and their relationship to risk for ovarian cancer. Cancer Epidemiol Biomarkers Prev 14:1125–1131

Salm SN, Burger PE, Coetzee S, Goto K, Moscatelli D, Wilson EL 2005 TGFβ maintains dormancy of prostatic stem cells in the proximal region of ducts. J Cell Biol 170:81–90

Summing-up

Siamon Gordon

Sir William Dunn School of Pathology, University of Oxford, South Parks Road, Oxford OX1 3RE, UK

I think we have had a feast during this symposium. I want to outline a few reminders of things that I have learned during this meeting, and things we might need to think of in terms of the subject as a whole. Today we have heard about chronic granulomatous disease and cystic fibrosis, as well as hyper-inflammatory states. There was some feeling that because we are dealing with environmental disease, genetic analysis will be less helpful. I think this is wrong. Even if these genetic disorders are rare, and even if they are masked by top-heavy environmental influences, they teach us so much that we must look for them and exploit them. An example is David Speert's work on signal transduction defects. This is the way we learn about important pathways.

Environmental factors are highly complex. We heard, for example, about how complex the constituents and effects of cigarette smoke might be. Richard Doll died recently, and it has taken 50 years to follow up some of those original population cohorts. There is still a big impact on society and it has taken a long time for the epidemiologists to persuade people to change their behaviour. It isn't easy to study what cigarette smoke is doing. Eric Bateman reminded us of the interplay of tuberculosis, dust, HIV and cigarette smoke. It is difficult to study this experimentally, and it is also not fashionable for experimental scientists. You don't get grants and high profile publications from this sort of work.

We need to start thinking of local effects. I remember Adrian Hayday saying that all politics is local and all disease has to take into account local circumstances. We learned beautifully about mucus. We should think more about how we stay healthy, and not only how we succumb to infection. We haven't given enough attention to the epithelium. Epithelial cells are quite heterogeneous. We heard a bit about goblet cells, but heard virtually nothing about Type 2 alveolar cells, which I think are important targets as well as producing surfactants. We heard a bit about the importance of epithelial cell polarity. We paid almost no attention to the lymphocyte populations that are intimately part of the epithelium, except for the dendritic cell. We don't know much about their growth or turnover, or about mutations within these cells. Epithelial cells not only have a barrier function, but also are capable of mounting an innate response, producing cytokines and mediators. They interact with white cells, and this is not always appreciated.

I said at the beginning that there are a couple of things that I would like to try to track during the meeting. Oxygen is one of these. The only way we discussed this was in relation to Anthony Segal's story about microbial killing by neutrophils, and the role of the oxidative metabolites. I still don't understand what happens when tissues become hypoxic, and when oxygen is toxic: are there levels where hypoxia could contribute to innate immunity? We have never discussed the real gas exchange function of the lung, except indirectly. We shouldn't forget about this.

We have talked a lot about host–microbe interactions, and in terms of the microorganisms themselves we know very little about how they attach, how they invade and the importance of the route of entry. What about a possible gastrointestinal route to lung infection? How do organisms get back into the airway if they do? Normally we think of airway going into the stomach rather than the reverse direction. Is it via a common compartment? Again, organisms are much more heterogeneous than we have discussed. Bacteria differ from one another and even within the same organism there is variation. There are virus differences, too. But there are also common elements. We have heard about some differences among fungi, which clearly have a lot to do with virulence factors and so on that make for unique as well as common features of response. Then we heard about coinfection with multiple strains of mycobacteria in the same patient. There is the hint that virus–bacterial interactions may be more selective. Also, it may depend on the sequence of infection and whether the virus or bacterium comes first.

An issue that we didn't address is the problem of dormancy or latency in tuberculosis. We heard about reinfection, but this miraculous property of a few organisms to survive dormant must be a tough thing to study. I was taken with Valerie Quesniaux's point about how these organisms do change. They have complex effects that are both pro- and anti-inflammatory. The sugar structures and acyl chains are changeable. These are not static organisms: they are dynamic and are very clever pathogens.

The effects of injury and repair on the resolution of inflammation and infection are important. I'm thinking of chronic obstructive pulmonary disease (COPD) as a major complication. We haven't tackled this at all. The question of going from an acute to a chronic inflammatory or infectious process, and then repair and scarring, is a major problem that we don't understand.

Are diseases like avian influenza diseases of innate immunity? I am not sure. They are fast and the outcome is determined quickly, so if anyone dies within a week of infection that is innate immunity (or lack of it). But this is still a vague, general concept. Then of course there is the impact of the delayed adaptive response. We know quite a bit about dendritic cells and how they can be manipulated, so that they become 'better than nature'. I'm not sure that this is true.

It is shameful that we know so little about the impact of nutrition on health and infection. It is an unfashionable topic, but we should pay more attention to it, especially in South Africa.

It is inevitable that in our research we are like Einstein's story of the man looking for his keys under the lamppost, not because that is where he lost them but because that is where the light is. We have systems we can study and techniques to study them, but we are sometimes guilty of not going where the questions are. Remembering this, we have heard about some nice mouse models and knockouts. One experiment that can be done in mice is adoptive transfer of bone marrow cells to ask whether we are dealing with a haematopoietic or a local environmental influence. We must remember that these models are quite damaging, to permit engraftment. Animals become sensitive to infection and the system is quite artificial.

The issue of cell trafficking is easier studied in animal models than humans. Traffic to the lymph nodes is really important. We also need to trace the organisms in the airway. Where do they go? Some of the methods to follow fluorescent bacteria in the intact animal aren't at the level of sensitivity we need at the moment, but non-invasive detection will be feasible in the future.

I keep wondering why we don't do better in making use of all the human material that is accessible. Can we detect organisms in the airway by looking at the composition of exhaled air? Are some of the modern methods of chemical detection sensitive enough? It would be good to look at infection in a non-invasive way.

One of the themes we have seen in several systems is that there are resident cells, and that new cells can be recruited from the blood. Neutrophils are mostly recruited, but they can remain within the lung. With mast cells and macrophages there are two populations, and this is probably true for other leukocytes. We need to be able to study *in situ*, not only where things are but, for example, where products get adsorbed to surfaces of other cells so you never know whether it is made by those cells or not.

If we go outside the intact organism, there is the possibility of primary and organ culture. You can recover macrophages relatively easily, but then you worry because once they are in culture the environment has changed. Primary epithelial tissue is difficult to study *in vitro* becoming more and more remote from real life. We need more and better systems for that.

We learned about cells: granules in neutrophils and eosinophils. We heard that receptors are not only Toll-like but there are Toll-independent systems. We heard about mediators, but perhaps we stressed the role of cytokines and chemokines too much. We didn't talk much about the low molecular weight metabolites, such as leukotrienes and prostaglandins, let alone the signalling pathways and their complexities. One of the target cells is the endothelium. This is what is causing death in many of these major virus infections. It would be interesting to know what the molecular targets are.

Another question I flagged in the introduction was can we use the blood to tell us what is going on in the lungs? We have heard the pentraxin story, but the problem with PTX3 is that it is fairly non-specific. Is it like discovering that you have a thermometer and you can measure temperature?

What are some of the issues that are still not easily explained? These include inflammation versus protection, apoptosis versus necrotic mechanisms of death. I wonder whether fungal infection is a sign of secondary immunodeficiency, or whether the infection itself is a cause of some of the problems that people heavily infected with systemic fungal burdens have. Do they have the fungus because they are sick or vice versa?

What about intervention? Perhaps PTX3 or maybe mannose-binding lectin (MBL) will be useful. The way these molecules discriminate between host and microbe is a fundamental problem. We have examples from these humoral molecules that there are cellular receptors. In the case of apoptotic cell recognition, we don't know which if any of those receptors are responsible for antigen presentation. This is a real gap in our knowledge. Anthony Segal reminded us that we don't know much about killing mechanisms. How do we know that the neutrophil isn't killing tubercle bacilli? In general, we don't know much about killing mechanisms within a 3D microenvironment in the host. Capture and clearance of organisms is not the same as killing. I was struck by the idea that shedding of epithelial cells is a neat way of getting rid of a whole package of dying cells and virus.

Immunotherapy is promising in other fields of chronic immunity. Tumour necrosis factor antibodies have been very useful in rheumatoid arthritis and Crohn's disease, but I don't think it offers much scope here. It may indeed be dangerous in reactivating dormant tuberculosis. We didn't say much about pharmacology and identification of novel drug targets. My last point is that we are dealing with biomedicine but we have social and political issues in the mix as well so that disease prevention and support of patients are paramount.

Contributor index

Non-participating co-authors are indicated by asterisks. Entries in bold indicate papers; other entries refer to discussion contributions.

Subject index